THE NUCLEAR POWER DECEPTION

About the Authors

ARJUN MAKHIJANI is President of the Institute for Energy and Environmental Research in Takoma Park, Maryland. He earned his Ph.D. in engineering at the University of California-Berkeley in 1972, specializing in controlled nuclear fusion.

A recognized authority on energy issues and nuclear issues in particular, Dr. Makhijani is the author or co-author of numerous reports and books in topics such as radioactive waste storage and disposal, nuclear testing, disposition of fissile materials, energy efficiency, and ozone depletion. He is the principal editor of *Nuclear Wastelands: A Global Guide to Nuclear Weapons Production and Its Health and Environmental Effects*, published by MIT Press in July 1995, and subsequently nominated for a Pulitzer Prize.

Dr. Makhijani has provided technical support to community activists on a variety of environmental and energy issues. He has also served as a consultant to United Nations agencies, many non-governmental organizations, and other institutions.

SCOTT SALESKA is a post-doctoral fellow at Harvard University's Department of Earth and Planetary Sciences. He has a B.S. in physics from MIT, and earned his Ph.D. in Energy and Resources at the University of California-Berkeley in 1998, specializing in ecological impacts of global climate change.

Dr. Saleska is author and co-author of numerous studies on radioactive waste disposal and environmental problems at nuclear weapons production facilities, as well as of scientific studies of ecological responses to global climate change. He worked at IEER as a staff scientist and consultant from 1990 to 1993.

THE NUCLEAR POWER DECEPTION

U.S. Nuclear Mythology from Electricity "Too Cheap to Meter" to "Inherently Safe" Reactors

by
Arjun Makhijani
and
Scott Saleska

A Report of the Institute for Energy and Environmental Research

The Apex Press
New York

Copyright © 1999 by the Institute for Energy
and Environmental Research

Published by The Apex Press, an imprint of the Council on International and Public Affairs, Suite 3C, 777 United Nations Plaza, New York, NY 10017 (800-316-2739)

Library of Congress Cataloging-in-Publication Data

Makhijani, Arjun
    The nuclear power deception : U.S. nuclear mythology from electricity "too cheap to meter" to "inherently safe" reactors / by Arjun Makhijani and Scott Saleska.
        p.   cm.
    "A report of the Institute for Energy and Environmental Research."
    Includes bibliographical references and index.
    ISBN 0-945257-75-9 (hardcover : alk. paper). —ISBN 0-945257-92-9 (softcover : alk. paper).
    1. Nuclear power plants—Risk assessment. 2. Nuclear industry—Environmental aspects. 3. Nuclear industry—Political aspects—United States. 4. Propaganda—United States. I. Saleska, Scott. II. Institute for Energy and Environmental Research (Takoma Park, MD). III. Title.
    TK9152.16.M33   1999   98-36673
    363.17'99'0973—dc21   CIP

Cover photo of "Atoms for Peace" mobile exhibit, U.S. Atomic Energy Commission, 1957. Under the program, initiated in the Eisenhower administration, the U.S. supplied highly enriched uranium for foreign research reactors in 41 countries. (U.S. Department of Energy)

To

W. H. Ferry,

visionary and mentor

CONTENTS

List of Figures and Tables   xi

Preface   xiii

Acknowledgments   xvii

Introduction: Summary and Recommendations   1

    Main Findings   1

    Recommendations   11

      Other Recommendations   12

PART ONE: HISTORY—NUCLEAR POWER PROPAGANDA AND REALITY

Chapter 1: Romance with the Atom   17

Chapter 2: Electricity Production and Nuclear Reactors   26

    Nuclear Fuel   30

      1. Uranium Fuel   30

      2. Plutonium Fuel   33

    Nuclear Reactors   36

      1. Thermal Reactors   37

      2. Breeder Reactors (Fast Neutron Reactors)   44

    The "Nuclear Fuel Cycle"   50

Chapter 3: The Early Years—Atomic Messiahs,
Propagandists, and Skeptics   53

    Atomic Messiahs and Propagandists   53

      1. "Atoms for Peace"   57

    Atomic Skeptics   62

      1. Early Practical Assessments   63

Chapter 4: Plutonium, the Nuclear Navy, and
Nuclear Power Development   70

    Round One: Dual-Purpose Reactors   71

Round Two: Admiral Rickover, the Nuclear
  Navy, and the Light Water Reactor   73
  1. Propaganda Aspects   73
  2. Shippingport   76
Other Skeptics   81
Chapter 5: From "Too Cheap" to Bust   84
The First Civilian Reactors   84
Safety   89
  1. Reactor Safety Basics   90
  2. Light Water Reactors: Basics about
  Loss-of-Coolant Accidents   92
  3. Historical Aspects of the Light Water
  Reactor Accident Debate   96
  4. Sodium-Cooled Fast Breeders   99
Cost   101
  1. The 1950s to the 1970s   101
  2. The 1980s and Early 1990s   104
Chapter 6: Radioactive Waste   108
Radioactive Waste Basics   109
  1. Waste Categories   111
  2. Quantities of Weapons-Usable Materials 114
High-Level Waste Management: Short-
  and Medium-Term Issues   115
  1. Nuclear Waste and Nuclear Power   119
  2. Reprocessing   121
Long-Term Management Issues   126
Concluding Observations   128

**PART TWO: MORE NUCLEAR POWER PLANTS OR A
SOUND ENERGY POLICY?**

Chapter 7: "Inherently Safe" Reactors—Commercial
  Nuclear Power's Second Generation?   133
New Reactor Designs: An Overview   135

New Light Water Reactor Designs 138

Other Reactor Designs 139

    1. The MHTGR 140

    Background 141

    Safety Concerns 143

The Semantics of "Inherent Safety" 148

Accidents and Nuclear Technology 150

    1. Chernobyl 153

    Health Effects 159

    2. Some Lessons of the Chernobyl Disaster 163

Chapter 8: Plutonium Disposition, Military Tritium,
and Commercial Reactors 165

Plutonium Disposition 166

Military Tritium 171

Concluding Observations on the Civilian-
Military Nuclear Power Connection 173

Chapter 9: Nuclear Power and Energy Policy 176

Energy Concepts 181

Renewable Energy Sources 184

Integrating Renewables with Energy Efficiency
in the Electricity Sector 189

    1. Obstacles to Efficiency 189

    2. A Proposal for a Financing Structure 192

    3. Improving Efficiency in Existing
Large Buildings 193

    4. Improving Efficiency in Existing
Small Buildings 194

    5. Other Measures 194

Combined-Cycle Plants, Nuclear Power Plants,
and CO_2 Emissions 195

The Kyoto Protocol 200

    1. Major Provisions of the Kyoto Protocol 201

    2. Joint Implementation 203

Appendix A: Basics of Nuclear Physics and Fission   207
    Structure of the Atom   207
    Radioactive Decay   208
    Binding Energy   210
    Nuclear Fission   211
    Fertile Materials   214
Appendix B: Uranium   215
    The Mining and Milling Process   217
    Conversion and Enrichment   219
    Regulations in the U.S.   220
    The Future   221
Appendix C: Plutonium   222
    Nuclear Properties of Plutonium   222
    Chemical Properties and Hazards of Plutonium   224
    Important Plutonium Compounds and Their Uses   225
    Formation and Grades of Plutonium-239   225

Glossary   229
References   241
Index   259

LIST OF FIGURES AND TABLES

Figure 1: Basic Elements of a Nuclear Power
Plant with a Separate Steam Generator   27

Figure 2: Converting Uranium into Low
Enriched Fuel for Nuclear Reactors   32

Figure 3: A Boiling Water Reactor (BWR)   41

Figure 4: A CANDU Type of Heavy Water
Power Reactor   41

Figure 5: A Schematic Diagram of a
Liquid Sodium-Cooled Fast Breeder Reactor   49

Figure 6: Nuclear Power Costs   102

Figure 7: Radioactivity and Thermal Heat Generation—
One Metric Ton of PWR Fuel (Heavy Metal Basis)   110

Figure 8: Contaminated Areas Around Chernobyl   157

Figure 9: Combined-Cycle Plant   198

Figure A-1: Distribution of Atomic Numbers of
Fission Products   213

Figure B-1: Uranium Decay Chain
(Half-lives Are Rounded)   216

Table 1: Composition of Fresh Enriched Uranium
Fuel and Spent Fuel   33

Table 2: Basic Characteristics of Reactor Types   46

Table 3: Chronology Leading to the First
U.S. Power-Generating Reactor   78

Table 4: Volume and Radioactivity of Wastes
in the United States by End of 1994   111

Table 5: Some Radioactive Waste Characteristics   114

Table 6: Historical World Plutonium Inventories
(in Metric Tons)   116

(in Metric Tons) 116

Table 7: Some Reactor Accidents 152

Table 8: Rejected Minimized Accessibility
Plutonium Disposition Options 170

Table 9: Comparison of Vitrification Options 172

Table 10: Stockpile Levels (1,000 and above)
and Tritium Requirements 174

Table 11: Energy Consumption and Electricity
Generation in the United States, 1992 177

Table 12: Renewable Intensive Global Electricity
Supply Projection (Billion Kilowatt Hours per Year) 186

Table 13: Comparison of Carbon Dioxide
Reductions: Natural Gas Combined-Cycle
versus Nuclear Power Stations 196

Table 14: Annex I and Annex B Parties 202

Table B-1: Uranium Isotopes and Natural
Uranium Composition 215

Table C-1: Physical Characteristics of
Plutonium Metal 222

Table C-2: Radiological Properties of
Important Plutonium Isotopes 224

Table C-3: How Plutonium Reacts in Air 225

Table C-4: Grades of Plutonium 226

PREFACE

In recent years, there has been a debate about the potential and need for developing a second generation of commercial nuclear power plants to generate electricity. Proponents of such development cite a range of reasons for undertaking it, primary among them the growing environmental problems (most notably the threat of global climate change) associated with conventional fossil-based electric power generation and the need to reduce the dependence of the United States on imported oil.

At the same time, a related debate is taking place about U.S. proposals to build one or more reactors for military-related purposes. The stated reasons for building new reactors have varied, ranging from new plutonium and tritium production reactors in the late 1980s and early 1990s, to reactors for burning excess military plutonium to a "triple play" reactor that would simultaneously burn excess plutonium, produce tritium (a radioactive gas used in nuclear warheads), and generate electricity. During the 1990s, a new element has been added to these debates—that of using new reactors to burn excess weapons plutonium.

At times, the two debates have converged, but not primarily for technical reasons. When political pressures to spend more money on reactors have been stronger, technical considerations have tended to take a back seat. When fiscal concerns have the upper hand, funds for military enterprises that would subsidize civilian power projects tend to be reduced or eliminated.

Before accepting arguments that nuclear power can alleviate the build-up of greenhouse gases or that joining military to civilian nuclear ventures is desirable, we need to learn what history might have to offer by the way of lessons. In particular, the idea of new reactors that would join military and civilian goals parallels the development of the first generation of power reactors in the United States. This study critically examines the history of wildly optimistic public statements that were made about nuclear power in the years and decades immediately following World War II and serves as a partial guide to dealing with critical civilian and military nuclear issues today. So far as we are

xiii

aware, the technical foundation of those extravagant promises has never been carefully scrutinized until now.

In 1954, Lewis Strauss, chairman of the U.S. Atomic Energy Commission, proclaimed that the development of nuclear energy would herald a new age. "It is not too much to expect that our children will enjoy in their homes electrical energy too cheap to meter," he declared to a science writers' convention (Strauss 1954). The speech gave the nuclear power industry a memorable phrase to be identified with, but also it saddled it with a promise that was essentially impossible to fulfill.

In contrast to the rosy propaganda and promises, commercial nuclear power from new nuclear plants has become the most expensive form of commonly used baseload electric power in the United States. In part, this was because utilities canceled 121 reactors in the post-1974 period; the money squandered on these canceled plants alone was about $44.4 billion in 1990 dollars (Komanoff and Roelofs 1992, p. 23), or about $50 billion in 1995 dollars. Even larger costs were incurred in the form of higher electricity costs for instance, because of the very high costs of plants completed in the 1980s. Enjoying virtually every conceivable advantage at its birth—from high public popularity to lavish government funding to virtually unanimous political support—the commercial nuclear power industry in the United States is a moribund one, with virtually every one of its early advantages reversed.

Part One of this study contains an introduction to the technical issues and then provides an historical analysis of nuclear power in the United States. In particular, it looks closely at the early claims that nuclear electricity would be "too cheap to meter" and whether they were, at the time, actually believed by the nuclear power proponents.

Part Two, drawing on the historical analysis, provides a critical appraisal of current plans for a second generation of nuclear plants partially subsidized by military materials production activities for nuclear weapons. It also reviews the persistent dangers in light of the Chernobyl accident and proliferation and environmental issues arising from the huge and growing stockpiles of weapons-usable plutonium in reactor spent fuel. This is followed by a chapter outlining an approach to creating an environmentally sound, reliable electricity system. There are also three appendixes: Appendix A on the basics of nuclear physics and fission, Appendix B on uranium, and Appendix C on plutonium. An introductory summary and recommendations is provided at the start of this book.

A note about sources: We have used original documents and sources for much of the material. Where the book covers ground that has already been covered by others, we have also used published books and materials as cited in these works. We have also used official historical accounts of the development of nuclear energy and of the history of the Atomic Energy Commission. For basic nuclear engineering information, we have used the textbooks, *Introduction to Nuclear Engineering* by John R. Lamarsh and *Nuclear Chemical Engineering* by Manson Benedict, Thomas H. Pigford, and Hans Wolfgang Levi. Unless otherwise stated, statistics on electricity costs, energy supply, and energy use are derived from the *Historical Statistics of United States from Colonial Times to 1970* (Washington, D.C.: U.S. Department of Commerce, 1975) and from various issues of the *Statistical Abstract of the United States*. Units are metric, unless otherwise noted. We have referred to literature produced by the Nuclear Regulatory Commission, generally abbreviated as NRC, as well as the National Research Council of the National Academy of Sciences, also generally abbreviated as NRC. In order to avoid confusion between these acronyms, we used the acronym NRC-NAS for the latter, and the acronym NAS to refer to studies by the NAS committees.

This book was originally issued in April 1996 as a report of the Institute for Energy and Environmental Research. It is being published as a book in English, French, and Russian in 1999. We have added sections in Chapter 9, comparing combined-cycle natural gas and nuclear power plants in terms of their ability to reduce carbon dioxide emissions, and on the Kyoto Protocol. Some copy editing changes have also been made. The new sections in Chapter 9 are excerpted from IEER's newsletter, *Science for Democratic Action*, Volume 6, Number 3, March 1998.

Arjun Makhijani
Takoma Park
November 1998

ACKNOWLEDGMENTS

I would like to thank Rolf Loschek and Margaret Hawley who did much of the early documentary research for this project. As with other Institute for Energy and Environmental Research (IEER) efforts, our librarian Lois Chalmers ably supported the research effort. Betsy Thurlow-Shields, Diana Kohn, and Pat Ortmeyer proofread the manuscript and made editorial comments. Julie Barnet provided extensive editorial suggestions on an early draft. Robert Alvarez, Kathleen Tucker, and Gar Smith provided comments on that draft which were very helpful in shaping the final work.

I am deeply indebted to Professor Lawrence Lidsky, Department of Nuclear Engineering, Massachusetts Institute of Technology, for his technical review of the manuscript of the first version of the report published in 1996. I am also very grateful to Dr. Charles Komanoff of Komanoff Energy Associates for his review and for the materials on economic aspects of nuclear power, and to Robert H. Williams, senior research scientist at Princeton University's Center for Energy and Environmental Studies, for materials on alternative energy futures. Of course, only the authors are responsible for any omissions as well as for the contents, conclusions, recommendations, and any errors that remain in this work. I would like to thank Ward Morehouse and the rest of the staff and board of The Apex Press, which has published other IEER works previously, for their continued confidence in our work. We would especially like to thank Cynthia Morehouse for her copy editing of the final product.

Principal funding for this work was provided by a special grant from the C.S. Fund. A portion of the funds for this book also came from general support grants from the Public Welfare Foundation, anonymous donor(s) via the Rockefeller Financial Services, and the Stewart R. Mott Charitable Trust. We are grateful for their support. I would also like to thank Gale M. Colby, Pushpa Mehta, Sunder Mehta, Anant Mehta, and Sanjay Mehta for their generous private contributions to IEER that aided the completion of this study.

Outreach work on *The Nuclear Power Deception*, including its publication in Russian and French, is a part of IEER's program on global nuclear material dangers, related energy and environmental issues, and nuclear disarmament. This program is generously funded by grants from the W. Alton Jones Foundation, the John D. and Catherine T. MacArthur Foundation, the C.S. Fund, the HKH Foundation, the New Land Foundation, and general support grants from the DJB Foundation and the Turner Foundation.

Arjun Makhijani
November 1998

INTRODUCTION: SUMMARY AND RECOMMENDATIONS

The global threats arising from nuclear power and large-scale fossil fuel use stem from a common failing in the political and economic structure of decision-making. Whether one considers plutonium or carbon dioxide emissions, there has been a consistent failure to ask and pursue vigorously the answers to a few simple questions before large-scale deployment of new technologies: Is there a potential for irreversible catastrophic damage if the system does not work? What is the fate of the most dangerous materials? How many generations could be affected?

One likely reason for the failure to pursue the answers even when the questions were asked is that power and money lies in the direction of development of these technologies, while the common good lies in the answers to the questions about damage to the Earth's ecosystems and to global security. But if we do not answer such questions, then it is possible that wide-ranging, potentially irreversible effects will damage the common good now and for uncountable generations into the future, even as individuals derive transient benefits from the technologies. This study examines nuclear power technology with such questions in mind.

Main Findings

1. *There was no scientific or engineering foundation for the claims made in the 1940s and 1950s that nuclear power would be so cheap that it would lead the way to a world of unprecedented material abundance. On the contrary, official studies of the time were pessimistic about the economic viability of nuclear power, in stark contrast to many official public statements.*

Unduly optimistic official public pronouncements regarding the promise of nuclear energy are epitomized by AEC Chairman Lewis

1

Strauss's forecast for an atomic age of peace and plenty in a 1954 speech, in which he also made his well-remembered remark:

> It is not too much to expect that our children will enjoy in their homes electrical energy too cheap to meter. . . .[1]

Our review of technical studies on nuclear power prepared in the 1940s and 1950s by a variety of government and industry sources, including the Atomic Energy Commission, revealed no evaluation of nuclear energy that concluded that nuclear energy would be cheap in the near future. On the contrary, many studies concluded that nuclear power would be more expensive than coal-fired electric generating stations and that it would have to be subsidized by military plutonium production in order to be economically viable. Other studies concluded that nuclear electricity might one day be competitive with coal, especially if coal prices rose and nuclear fuel were cheap.

A 1948 AEC report to Congress cautioned against "unwarranted optimism" about nuclear power because there were many "technical difficulties" facing it that would require time to overcome.[2] The report's authors included many of the leading nuclear scientists of the time, including Enrico Fermi, Glenn Seaborg, and J.R. Oppenheimer.

As another example, an industry report done by four industry-utility groups, including Bechtel, Monsanto, Dow Chemical, Pacific Gas and Electric, Detroit Edison, and Commonwealth Edison, concluded in 1953 that "no reactor could be constructed in the very near future which would be economic on the basis of power generation alone."[3] Ward Davidson, a research engineer with Consolidated Edison, one of the country's largest utilities, laid out in 1950 the technical difficulties facing practical nuclear power in considerable detail, including problems such as making durable materials of assured quality that could survive the intense neutron bombardment in nuclear reactors.

Insiders even scorned those who might suggest that nuclear energy would usher in an era of plenty, as noted by C.G. Suits, a General Electric Vice-President, in December 1950:

> It is safe to say . . . that atomic power is *not* the means by which

---

[1] Strauss 1954.

[2] AEC 1948, p. 43.

[3] *Nucleonics* 1953, p. 49.

man will for the first time emancipate himself economically
. . . . Loud guffaws could be heard from some of the laboratories
working on this problem if anyone should in an unfortunate
moment refer to the atom as the means for throwing off man's
mantle of toil. It is certainly not that! . . . This is expensive
power, not cheap power as the public has been led to believe.[4]

2. *Cold War propaganda rather than economic reasoning was a driving force behind the rush to build a commercial nuclear power plant in the United States.*

Coal was plentiful in the United States, then as now. Further, there were no serious pressures in the 1950s to reduce its use due to the build-up of greenhouse gases. There was therefore no urgent reason then to press ahead with building large nuclear power plants.

Despite the pessimism of governmental as well as corporate studies about its economics, nuclear electricity was seen by many in government as a key technology to be exploited in the Cold War with the Soviet Union. Atomic Energy Commissioner Thomas Murray said in 1953 that peaceful applications of the power of the atom "increases the propaganda capital" of the U.S. relative to the Soviet Union.[5]

The chairman of the congressional Joint Committee on Atomic Energy warned in 1953 that "the relations of the United States with every other country in world could be seriously damaged if Russia were to build an atomic power station for peacetime use ahead of us. The possibility that Russia might demonstrate her 'peaceful' intentions in the field of atomic energy while we are still concentrating on atomic weapons, could be a major blow to our position in the world."[6] President Eisenhower's 1953 "Atoms for Peace" speech was perhaps the centerpiece of the U.S. effort to cast its nuclear program in a peaceful light for the purposes of Cold War propaganda.

The choice of the light water reactor (LWR) as the first commercial reactor was influenced by these Cold War considerations. H.C. Ott of the AEC's division of reactor development objected to the selection of the pressurized water reactor (PWR) in 1953 because the choice had

---

[4] Suits 1951. The speech was given in December 1950 and printed in February 1951 in *Nucleonics*. (Paragraph breaks not shown here; for paragraph breaks in the quote, see Chapter 3, pp. 62-63.)

[5] Murray 1953c.

[6] Cole 1953.

not been justified "as a logical part of the overall reactor development program and no arguments are advanced to support the thesis that a prototype power plant should be built."[7] But a classified 1954 AEC report concluded that the PWR had the best short-term prospects. However, the design was considered a poor long-term choice because it was (erroneously) thought at the time that the scarcity of uranium would make reactors like PWRs uneconomical because they were net consumers of fissile material. The actual vulnerabilities turned out to be not uranium-235 scarcity, but safety, capital cost, and proliferation.

Solar energy was not pursued seriously despite a 1952 report of a presidential commission (the Paley Commission) that anticipated oil shortages in the 1970s. The report was relatively pessimistic about nuclear energy and called for "aggressive research in the whole field of solar energy—an effort in which the United States could make an immense contribution to the welfare of the world."[8] This advice was ignored. Solar energy research and development could not provide the Cold War "propaganda capital" that nuclear power gave to the U.S.

3. *The AEC overruled some of its own personnel and the official Advisory Committee on Reactor Safeguards (ACRS) in its rush to build large-scale power plants that would feed electricity into utility grids.*

The dangers of nuclear power plants were well enough understood in the 1950s that there were many voices advising against a rush to build large-scale plants. As early as the mid-1950s, the ACRS advised the AEC to proceed more slowly and cautiously with its licensing of a sodium-cooled breeder reactor near Detroit, Fermi I. Autoworkers' unions went all the way to the Supreme Court to try to block its construction, but they failed. A pilot plant of similar design, the Experimental Breeder Reactor I built in Idaho, had had an accident in 1955 during a safety experiment. The voices of caution did not prevail. The Fermi I reactor was started up in 1963, but had not yet achieved full power when it suffered a partial meltdown due to a partial cooling system blockage in 1966.

The ACRS also had concerns regarding unresolved issues surrounding a potential meltdown of the core of light water reactors

---

[7] Ott 1953.

[8] Paley Commission 1952, Vol. IV, p. 220.

(LWRs) in case of a loss-of-coolant accident. Its concerns received heightened attention only after protracted hearings in the early 1970s on the issue in which independent scientists, notably from the Union of Concerned Scientists, and whistleblowers played central roles.

Finally, in 1957, an official study by Brookhaven National Laboratory concluded that up to 3,400 people could die and up to 43,000 people could be injured in case of a severe accident in a 500 megawatt-thermal (100 to 200 megawatt-electrical) nuclear power plant. Property damage alone was estimated at up to $7 billion in 1957 dollars (about $38 billion in 1995 dollars). Many power reactor safety studies and experiments were begun in the 1950s, but they were not completed before commercial reactors started being built in large numbers. Instead the government provided industry with a huge subsidy in the form of the Price-Anderson Act, which limited their liability to $500 million above "private insurance." Self-insurance was permitted as a form of private insurance under the Act. The total compensation was therefore considerably less than 10 percent of the estimate in WASH-740 of property damage alone.

The potentially catastrophic nature of nuclear accidents was made painfully graphic by the Chernobyl accident. The amount of damage from a catastrophic accident on the scale of Chernobyl vastly exceeds the $7 billion insurance provision in the 1988 amendment to the Price-Anderson Act.

4. *Every major reactor design that was adopted had, and continues to have, crucial unresolved safety vulnerabilities as a result of the rush of the nuclear weapons states to deploy nuclear power plants well before the technology had been properly investigated and developed.*

The three major reactor designs of the 1950s were:

- Water-cooled and water-moderated reactors;

- Graphite-moderated reactors (gas-cooled or water-cooled);

- Sodium-cooled fast neutron reactors (also called breeder reactors).

The basic safety flaws of these designs were identified in the 1950s and 1960s. Yet nuclear establishments (governmental and private) persisted with these designs because they had already made heavy investments in them. One initial AEC response to safety problems as they were revealed was to try to suppress discussion. But in the United States, where the public's right to know is greater than in any other major nuclear state, many safety issues did become public. The official response in such cases was to try to deal with safety issues in an ad hoc and dilatory manner. When that route became too cumbersome and bred lawsuits, the AEC held a rule-making hearing on the most import-ant kind of accident identified for LWRs—the loss-of-coolant accident. The hearings revealed serious problems, official attempts to cover them up, and safety issues that had not yet been satisfactorily resolved. All this resulted in a loss of public confidence and higher costs in the long run due to retroactive safety requirements that were needed.

Basic safety issues have nòt been resolved in that the potential for catastrophic accidents remains. For instance, then-NRC Commissioner James Asselstine, noted in 1986 that "given the present level of safety being achieved by the operating nuclear power plants in this country, we can expect to see a core meltdown accident within the next 20 years, and it is possible that such an accident could result in off-site releases of radiation which are as large as, or larger than, the releases estimated to have occurred at Chernobyl."[9]

Newer versions of the light water reactors in the United States and elsewhere have not eliminated the danger arising from a loss-of-cool-ant accident. Similarly, safety vulnerabilities continue to exist in other reactor designs that are being advertised as "inherently safe," such as a gas-cooled, graphite-moderated reactor. This is a public relations phrase that applies to avoiding or minimizing risks from particular types of accidents, notably loss-of-coolant accidents. Other types of accidents, also potentially catastrophic, may occur with such designs.

There is an even greater danger of a severe accident in the former Soviet Union, resulting from relatively less safe designs (including lack of secondary containment) and poorer maintenance due to the economic collapse that has taken place at the end of the 1980s. Accord-ing to Michael Golay, a professor in the nuclear engineering depart-ment of the Massachusetts Institute of Technology, there is "a very high likelihood of a serious reactor accident in the near future" in the former Soviet Union and Eastern Europe due to the relatively unsafe reactor

---

[9] Asselstine 1986.

designs, poor reactor system construction materials, and other factors, such as low spending on maintenance.[10]

5. *Nuclear power became established in the marketplace at a low price in the 1960s as a result of government subsidies, lack of adequate attention to safety systems, and an early decision by manufacturers to take heavy losses on initial orders. Costs increased when these advantages were reduced.*

Direct government subsidies helped finance reactors until 1962. After that, G.E. and Westinghouse decided to market LWR technology at a loss because they feared it would otherwise become obsolete.

A General Electric vice-president later stated that:

> If we couldn't get orders out of the utility industry, with every tick of the clock it became progressively more likely that some competing technology would be developed that would supersede the economic viability of our own. Our people understood this was a game of massive stakes and that if we didn't force the utility industry to put those stations on line, we'd end up with nothing.[11]

About 45 percent of the entire eventually installed nuclear capacity of about 100,000 MWe in the United States was ordered in the four-year period between 1963 and 1967. There was a rush of reactor orders and nuclear power plant completions. The power plants were built with government-subsidized insurance and without practical assurance that critical safety systems would work.

Several factors led to nuclear power costs increasing rapidly in the 1970s and early 1980s:

- Manufacturers stopped offering the plants at large losses.

- Government subsidies (other than accident insurance under the Price-Anderson Act) disappeared.

- The AEC imposed new safety requirements.

- Interest rates rose dramatically in 1979 and stayed relatively

---

[10] Golay 1993, pp. 1-4.
[11] Quoted in Demaree 1970, p. 93.

high for over a decade.

As a result of these factors, nuclear power costs rose very rapidly. At the same time, electricity growth rates declined. As the 1980s wore on, nuclear power plant operating, maintenance, and fuel costs rose to above those for coal-fired plants, as did estimates for reactor decommissioning. As a result, the loss of public confidence in nuclear power spread from Main Street to Wall Street. The disillusionment was the result of promises on cost that could not be fulfilled, a poor approach to safety that ignored or downplayed early warnings, and attempted cover-ups of safety issues that were later exposed. The Three Mile Island accident was perhaps the knock-out punch on safety issues. While the releases of radioactivity were not high (relative to other severe nuclear accidents), it showed that core meltdown accidents could happen and that accident probabilities may not be as low as had been supposed. Public disillusionment with nuclear power was reinforced by a corresponding loss of trust on nuclear weapons-related environmental mismanagement and misrepresentations.

6. *The non-proliferation issues related to nuclear power, and in particular their relation to the arsenals of the existing nuclear weapons states, have never been satisfactorily resolved.*

Nuclear power is vulnerable to forces of social instability and violence, which are becoming more technologically sophisticated while crucial institutional mechanisms for holding them in check remain weak. Examples range from the bombing in Oklahoma City to the gas attack on the Tokyo subway to threats of radioactive warfare made by Chechen rebels.

A second proliferation vulnerability of nuclear power is the vast amount of plutonium created in nuclear power plants. Every four years or so commercial nuclear power reactors create an amount of plutonium equal to that in the global military stockpile. This plutonium is not usable for weapons unless the spent reactor fuel is reprocessed (that is, unless the plutonium is separated from fission products and residual uranium). A large amount of plutonium has already been separated. Commercial reprocessing plants are operating in France, Britain, Japan, Russia, and India. Proposals exist for reprocessing U.S. spent fuel more (see below).

7. *Management of spent nuclear fuel has become a central concern*

regarding nuclear power growth.

The problem of high-level nuclear waste has not been resolved anywhere in the world, after four decades of nuclear electricity generation. The early confidence that nuclear scientists would somehow solve the waste problem, just as they had built the atom bomb, has evaporated.

In the United States, sound science has been overtaken by political considerations in the rush to relieve utilities of the liabilities deriving from the spent fuel accumulating at reactor sites. The Yucca Mountain site is the only one under investigation. Yet calculations done for this site, including those by Department of Energy contractors, indicate that radiation doses could be hundreds or even thousands of times above presently allowable limits. There is pressure to relax standards and change calculation methods instead of improving the repository program.

8. *Reprocessing, which is the separation of plutonium and uranium from used reactor fuel, is a costly, dangerous, and proliferation-prone technology. Yet political pressure is building to reprocess spent fuel as a waste management method.*

Reprocessing is very costly. The overall costs of this approach to spent fuel management and disposal would range from roughly $130 billion for government-subsidized reprocessing to $240 billion for commercial reprocessing in which the customer pays the full costs. This translates into half-a-cent to one cent per kilowatt hour of nuclear electricity—five to ten times more than the contribution of 0.1 cent per kilowatt hour (electrical) that nuclear utilities are now required to make by law into a nuclear waste fund. This presumes commercial reprocessing in a new reprocessing plant in the United States.

The safety and environmental consequences of using existing U.S. military reprocessing plants, which are over 40 years old, are incalculable. As a result, the costs are also highly uncertain. Reprocessing in existing military plants would greatly exacerbate high-level waste management problems, already beset by risks of fires and explosions.

Commercial reprocessing in France, Britain, Japan, Russia, and India is now the most important contributor by far to the growth of nuclear weapons-usable plutonium in the world. As the only nuclear weapons state not reprocessing for military or commercial reasons, the U.S. is the one country with the political standing to persuade Russia and other countries to stop commercial reprocessing. Reprocessing U.S. commercial spent fuel in the United States (or abroad) would be a grave practical setback to the implementation of the U.S. non-proliferation policy of discouraging reprocessing and the growth of weapons-usable materials stocks.

Yet the political pressure from the U.S. nuclear power industry is causing proposals for reprocessing this commercial spent fuel to reemerge in a way not seen since the early days of the Reagan administration. One such proposal was put forward by Westinghouse in an August 1995 study done for the DOE. This is a dangerous trend. It is difficult to overemphasize the central importance of U.S. policy against commercial reprocessing.

9. *The surplus plutonium in U.S. and Russian military stockpiles is exacerbating a growing commercial plutonium surplus.*

Proposals have been put forward to convert plutonium into mixed uranium-plutonium oxide fuel (called MOX fuel) in order to use it in existing commercial reactors or in new reactors. Such use would create a new commercial-military link that seems in many ways like a replay of the earlier debates on dual-use reactors. This time around, however, the civilian reactors are proposed to be used to partially destroy (or "burn") rather than create military plutonium. Such a connection would be undesirable for a number of reasons, not least of which are the proliferation implications of establishing a plutonium fuel economy in the United States.

10. *Nuclear power plants cannot simultaneously meet stringent safety criteria that would rule out catastrophic Chernobyl-like accidents and also contribute significantly to the reduction of greenhouse gas emissions in a timely manner.*

In order to ensure that nuclear reactors are not vulnerable to catastrophic accidents, new designs would need to be developed. These designs would have to be thoroughly checked on paper and in experimental and pilot-scale reactors *before* relatively large plants were built. Such an effort to ensure reactor safety and regain public confidence would take decades, if it can be accomplished at all. However, carbon dioxide emissions must be reduced in the same period. The next few decades will be crucial in the effort to minimize the threat of disastrous adverse effects due to the build-up of greenhouse gases. As a result, nuclear power plants cannot simultaneously meet stringent safety criteria that would rule out catastrophic Chernobyl-like accidents and also contribute significantly to alleviating the greenhouse gas build-up.

11. *It is possible to simultaneously phase out nuclear power plants and reduce carbon dioxide emissions from fossil fuel burning.*

The low efficiency of primary fuel use, even in technologically advanced countries and the even lower efficiencies in the rest of the world, means that much or most of the increased energy needs of the world's people can be met by improving efficiency dramatically. Renewable energy sources and judicious use of natural gas technologies can displace both nuclear power plants and help reduce the use of oil and coal in electricity generation. Technologies for greatly increasing energy efficiency as well as for using renewable energy sources are available now. Many others are nearly commercial. Institutional and market failures as well as the lack of proper government procurement policies and regulations are systematically hindering the widespread use of these technologies.

12. *For a given sum of money, combined-cycle natural gas plants reduce carbon dioxide emissions more than nuclear power plants.*

Natural gas burning results in carbon dioxide emissions at about half the level of coal per unit of heat energy produced. However, the efficiency of combined-cycle power plants using natural gas is so high and their cost so low relative to nuclear power plants that greater reductions in CO_2 emissions can be realized if combined-cycle plants are used to displace existing coal-fired power plants.

Recommendations

The central recommendation of this study is that nuclear power should be phased out. Specifically:

Existing nuclear power plants should be phased out as they come to the end of their licensed lives, or earlier if that is compatible with or needed for the security and safety of power supply.

New nuclear power plants should not be built in the foreseeable future.

The phase-out of nuclear power should be accomplished simultaneously with a reduction in emissions of carbon dioxide from fossil fuel use by greatly increasing energy efficiency, moving to renewable sources of energy as the primary energy supply, and using natural gas judiciously. The "Atoms for Peace" program, which is a dangerous relic of the Cold War, should be replaced by a global "Energy for Peace" program that stresses renewables and energy efficiency.[12]

Increasing the efficiency of energy use and increasing generation of electricity from renewables should be accomplished by policies, such as:

- Requiring developers of new buildings and factories to contribute to an electricity capital fund corresponding to the demand their project makes on the electricity grid.

- Enacting procurement policies by which governments and utilities would acquire electricity generated from renewable energy sources in order to encourage the development and widespread use of these technologies.

- Creating a "just-in-time" system for investments in small-scale cogeneration plants, small- and medium-scale renewable energy power plants, and energy efficiency to reduce the average amount of idle capacity.

Other Recommendations

1. Surplus plutonium should be vitrified rather than used as a reactor fuel. No infrastructure for use of mixed oxide fuel (a mixture of plutonium oxide and uranium oxide) should be created. Proposals to burn plutonium in existing commercial reactors or to produce military tritium in them should be scrapped.

2. Because military involvement in the development of commercial nuclear power has encouraged poor energy choices from the perspective of civilian power needs, this course should not be repeated. In particular, the DOE should halt any further consideration of building a new dual-purpose production reactor

---

[12] The phrase "Energy for Peace" was coined by George Perkovich of the W. Alton Jones Foundation.

which would both produce tritium for bombs and commercial electricity or a "triple play" reactor which would burn excess weapons plutonium in addition to these two functions. All expenditures for research and development along these lines should be stopped.

3. Industry must accept the financial risks of possible failure. This includes an end to federally established liability limits embodied in the Price-Anderson law for new nuclear power plants. Further, the goverment should collect fees for insurance for existing power plants that correspond to damage assessments that take into account the scale of harm inflicted by the Chernobyl accident.

4. Maintenance of a knowledge base regarding nuclear technology is important for a number of reasons, including medical and research uses, improving reactor safety so long as power reactors are in operation, and study of long-term waste management. This function should be performed openly by universities and other public and private research centers that are not connected to the secrecy prevailing in nuclear weapons establishments.

5. Management of spent fuel and weapons-usable fissile materials involves such momentous and unprecedented security and environmental issues over so many generations that it must be done in the most democratic, scientifically thorough manner of which society is capable. Management of wastes that already exists should be distinguished from waste that would be produced by new nuclear power plants. The government should not agree to simply take over the liabilities of nuclear waste from new plants.

The policies needed to restructure the existing waste management program are discussed in our earlier book on nuclear waste, *High-Level Dollars, Low-Level Sense.*[13] We merely summarize those recommendations and update them here:

• Reclassify waste by longevity and hazard.

---

[13] Makhijani and Saleska 1992.

- Cancel the Yucca Mountain repository program.

- Give due consideration to the historical claims of Native Americans to the land upon which Yucca Mountain is situated.

- Create on-site storage with adequate vigilance regarding safety of construction, siting issues, and electricity requirements, in the context of local public utility commission proceedings, as an interim measure while the long-term management program is restructured.

- Rule out reprocessing as a waste management option.

- Focus the long-term management program on the scientific issues that must be understood. Examples are:

    ⇒ study of natural conditions in which radioactivity does not migrate;

    ⇒ creation of engineered barriers to mimic natural materials that retard radionuclide migration;

    ⇒ development of methods to reduce uncertainties regarding impact on the human environment in the long-term;

    ⇒ assessment of non-repository options, such as sub-seabed disposal or disposal deep within the Earth's crust (which extends down to about 40 kilometers below the surface).

- Give the responsibility for long-term nuclear waste management to a new institution that is free of the conflict-of-interest that pervades the present program. DOE has been a major generator of high-level waste and is at the same time responsible for repository site characterization and selection.

PART ONE: HISTORY—
NUCLEAR POWER PROPAGANDA
AND REALITY

. . . you had uranium in the rocks, in principle, an inexhaustible source of energy—enough to keep you going for hundreds of millions of years. I got very, very excited about that, because here was an embodiment of a way to save mankind. I guess I acquired a little bit of the same spirit as the Ayatollah has at the moment.

—Alvin Weinberg, former head of the Oak Ridge
National Laboratory and nuclear
reactor designer, 1981[14]

I am sure we are agreed that the ultimate survival of America is dependent on intellectual vigor and on spiritual deeprooting—not on specific devices which are always for the moment. The atom has no ethics of its own any more than it has politics. The future of the scientists' America, and yours and mine, lies fundamentally with education—that which is taught to the young in our schools— that which is taught throughout life in the media of general communication by the comtemporary writers. Fundamental are respect and zeal for scholarship, a lively regard for moral values, and a love of truth. And of these the last is, of course, the greatest.

—Lewis Strauss, AEC Chairman, 1954[15]

---

[14] Alvin Weinberg in 1981 interview, quoted in Ford 1982, p. 25.
[15] Strauss 1954

CHAPTER 1:
ROMANCE WITH THE ATOM

All forms of transportation will be freed at once from the limits now put upon them by the weight of present fuels. . . .

Instead of filling the gasoline tank of your automobile two or three times a week, you will travel for a year on a pellet of atomic energy the size of a vitamin pill . . . The day is gone when nations will fight for oil. . . .

The world will go permanently off the gold standard once the era of Atomic Energy is in full swing . . . With the aid of atomic energy the scientists will be able to build a factory to manufacture gold.

No baseball game will be called off on account of rain in the Era of Atomic Energy. No airplane will by-pass an airport because of fog. No city will experience a winter traffic jam because of snow. Summer resorts will be able to guarantee the weather and artificial suns will make it as easy to grow corn and potatoes indoors as on a farm.

—David Dietz, science writer, 1945[16]

The control of fire was central to the development of cities—that is, of civilization. Its use on a large scale directly and indirectly, through diverse carriers of energy such as steam, is the foundation of modern industry and commerce. In the eighteenth and nineteenth centuries, steam power enabled the centralization of manufacture by the use of mechanized transport to draw huge quantities of raw materials, such as cotton and jute, as well as food from around the globe into the world's new manufacturing centers in Europe. The first fuel to be used on a large scale for industry and transport was coal; it continues to occupy a large place in the world's energy supplies today. In the late

---

[16] David Dietz, quoted in Ford 1982, pp. 30-31.

nineteenth century, it was joined by petroleum; in the twentieth, natural gas was added to the fossil fuel mix.

The potential for the application of energy to transform life for vast numbers of people was demonstrated in the second part of the nineteenth and first half of the twentieth centuries in a number of radical and graphic ways. Electric lights illuminated the night. Rapid travel over large distances became commonplace, first via railroad and steamships and then also via trolleys, buses, and cars. This mechanized mobility was symbolized by Phileas Fogg, Jules Verne's nineteenth-century fictional voyager, who went around the world in 80 days and returned home *punctually*. Farm mechanization reduced the need for farm labor; cities grew; occupations and specializations multiplied.

In the cities, automobiles and street cars rapidly replaced horse-drawn carriages. And the domestic scene was transformed, for those who could afford it, by central heating and numerous appliances that reduced the burdens of physical labor. The possibility that life for ordinary people around the world could one day be very comfortable, even luxurious, was no longer theoretical—it was being practically realized every day by large numbers of people of European origin and also by a small minority in the colonized countries. The prospect that such a life would be available to all seemed to depend on nothing more than human ingenuity in the application of science and technology and on the availability of sufficient natural resources, chief among which were fuels.

But these historic changes also carried the seeds of misery and destruction. Consolidation of farms threw people off the land. Machines threw people out of work. In many of the countries that were colonies of Europe, the destruction of cottage industries actually reduced the proportion of people working in non-agricultural occupations. At the same time, there was little work to be had on the land. From time to time, as in the 1930s in the United States, there were vast and sudden displacements of people from farms to urban areas, accelerating trends started by industrialization. Indigenous cultures, whose knowledge of the natural environment is still unparalleled by science in many ways, were destroyed around the world. Unemployment became a permanent feature of the world economy.

Despoilation of the environment was occurring on a scale as grand as the huge industries that were springing up. Air pollution was, in many places, literally breathtaking. For instance, in London the air often got so bad that episodes of smog came to be called "peasoupers" after their resemblance to pea soup: visibility was typically reduced to a few yards. Thousands of people died of respiratory diseases as a result of the London "peasouper" of 1952. The public outcry accompanying the deaths and suffering led to the initiation of unprecedented pollution control regulations in Britain. The general recognition of potential damage to the entire atmosphere due to a build-up of carbon dioxide from fossil fuel burning, however, was still about three decades away.[17]

As the exploitation of resources and the trade in them became global, so did the wars for their control. To a considerable extent, these global wars had their roots in the dependence of Western economies on cheap imported primary commodities and in the competition between them for these resources. After World War I, oil rapidly became the most crucial strategic primary commodity. Much of the prelude to World War II, including the Japanese bombing of Pearl Harbor, many of the battles during that war, and much of the wartime strategy of the antagonists revolved around the control of oil resources that had become the lifeblood of the war machine.[18]

By the middle of the twentieth century, with the colonies in Asia and Africa on the verge of political independence, people throughout the world were seeking to achieve the level of material standards of living that had already become a reality for a substantial minority of people in western Europe and the United States and would soon be realized by a majority. But would there be enough resources for all, given the already high and rising levels of consumption in Europe and the United States and the dependence of Western economies on import-

---

[17] Svante Arrhenius, a Swedish chemist, calculated in 1896 that a doubling of carbon dioxide levels could increase the Earth's temperature by 4 to 6 degrees Celsius. Even before that, in 1827, the eminent French mathematician and physicist, Jean Baptiste Fourier, warned that industrial activities may affect the Earth's climate. For a historical account of scientific investigations of climate change, see Falk and Brownlow 1989. However, the threat to the Earth's ecosystems and to society posed by the build-up of greenhouse gases, of which carbon dioxide is the most important one, did not become an important global policy issue until the 1980s.

[18] Yergin 1991 discusses these issues at length. See, for instance, pp. 308-323 for the prelude to Pearl Harbor. The United States moved its Pacific fleet from its southern California base to Pearl Harbor in 1940 partly to deter Japanese demands for more Indonesian oil.

ed primary commodities, especially oil?

Einstein's discovery early in the twentieth century that matter and energy were equivalent, expressed by the famous equation $E = mc^2$, came in the middle of this immense and unprecedented technological, political, economic, and military ferment. H.G. Wells, in *The War of the Worlds*, wrote about bombs that might destroy cities and entire civilizations. But there were also visions of unlimited amounts of energy for everyday life. Einstein's equation showed that a small amount of matter was theoretically equivalent to a huge amount of energy: just one gram of matter, if completely converted to energy, was equivalent to roughly 3,000 metric tons of coal.[19]

If only some way could be found to change matter into energy, the days of deprivation would be over! The Pharaohs needed slaves to do their bidding. Modern life would not need to be cruel to be affluent. Small bits of dead matter could take the place of slaves and everybody could be happy ever after—at least so far as material matters were concerned. Life would be free of drudgery. Convenience and creativity would flourish in the ample leisure time that everyone would enjoy.

In the late 1930s, the *fission* of uranium—that is, the splitting apart of its nucleus into smaller nuclei of light elements—was discovered and the possibility of converting matter into energy on a large scale started to move from the realm of science fiction and improbable theory to reality.

The practical harnessing of fission energy required the splitting of a large number of uranium atoms—in a controlled sustained way for nuclear power production, or all at once for a bomb. The Hungarian scientist Leo Szilard had realized well before fission was discovered in the laboratory in 1938 that a nuclear *chain reaction* would be the basis for nuclear energy production, whether commercial or military. In such a reaction, each fission would generate another without any external inputs so that, once initiated, fission reactions would continue until some other factor intervened to stop them.

---

[19] See Appendix A for details of how the calculation is done. The symbol E in Einstein's equation stands for energy, m for mass, and c for the speed of light. The symbol c^2, pronounced "c-squared," stands for the speed of light multiplied by itself. A gram is about one-thirtieth of an ounce. A pair of gold earring studs typically weighs a few grams. There are one million grams in a metric ton, which is about 10 percent larger than a U.S. ton of 2,000 pounds. A U.S. ton is also called a short ton.

Uranium appeared capable of sustaining a chain reaction because each fission released more than one neutron, a neutral particle that could penetrate the outer parts of an atom to reach its tiny nucleus. After the experimental demonstration of fission in Germany in late 1938 and its confirmation in the United States in 1939, the main question that remained was: could a nuclear chain reaction be realized in practice? If so, a large sustained release of energy could be achieved. The requirement for achieving a nuclear explosion was even more stringent since each fission would have to generate more than one fission in a very short time. In this way, the number of fissions would multiply very rapidly, resulting in a huge explosive release of energy.

The first chain reaction took place in an "atomic pile," as nuclear reactors were initially called, at the University of Chicago in December 1942. However, a minimum amount of nuclear material, called a *critical mass*, was necessary to sustain a chain reaction. The most basic physics questions had been answered. The immense engineering job of making nuclear energy a practical reality for explosive or commercial applications remained.

Commemorating the first self-sustaining chain reaction at the University of Chicago, on the occasion of the 5th Anniversary, December 2, 1947, are left to right: Atomic Energy Commissioners William W. Waymack and Robert F. Bacher, Farrington Daniels, Walter H. Zinn, Enrico Fermi, and R.M. Hutchins, chancellor of the University of Chicago. (Reprinted with permission of Argonne National Laboratory)

It was thought early on that the widespread use of nuclear energy would be complicated by an important resource limitation. While heavy elements could be fissioned to yield energy, only one element that occurred in nature in substantial quantities could sustain a chain reaction. That element was uranium. There was a further difficulty. It was discovered that only one naturally occurring isotope of uranium, called uranium-235, could sustain a chain reaction (see box). However, about 99.3 percent of natural uranium consists of uranium-238, which cannot sustain a chain reaction. Uranium-235 is only about 0.7 percent of natural uranium. Still, just one gram of uranium-235, when completely fissioned, yielded as much energy as three metric tons of coal, which is more than annual average household energy requirement for home heating in the United States.

Isotopes of Elements

Elements occur in variants called isotopes. All isotopes of an element have essentially identical chemical properties, which are determined by the number of protons in the nuclei of the element's atoms. Protons have positive electrical charges. The number of protons in a nucleus is normally equal, at ordinary temperatures, to the number of electrons that surround the nucleus in that atom. But the nuclei of elements can also contain varying numbers of neutrons, which are electrically neutral particles slightly heavier than protons, and far heavier than electrons. Changing the numbers of neutrons in the nucleus changes the properties of the nucleus and the overall weight of the atoms of an element. Variants of an element whose nuclei have the same number of protons but different numbers of neutrons are called isotopes of that element.

Some heavy nuclei are rendered highly unstable and split apart after absorbing a slow neutron, having essentially no kinetic energy. Such isotopes are said to be fissile. Other heavy nuclei require incoming neutrons (or other particles) to have a large amount of energy before they will split apart. These isotopes are fissionable but not fissile. In general, fissile isotopes are required to sustain chain reactions, and hence to build nuclear reactors or nuclear weapons. Uranium-235 is essentially the only naturally occurring fissile material.[20] Uranium-238 is fissionable but not fissile.

---

[20] Trace amounts of plutonium occur in nature but they are of no practical significance.

But uranium-238 was soon found to possess another remarkable property that made it seem at least as important a substance as uranium-235. When uranium-238 absorbs a neutron, it is transmuted, in two steps, to a fissile element that is present in nature only in the minutest quantities: plutonium-239 (see Appendix A). This meant that a nuclear reactor could be used to do two things at once. First, it could generate energy by fissioning uranium-235 in a chain reaction. Second, it could at the same time convert non-fissile uranium-238, which was 140 times more plentiful than uranium-235, into fissile plutonium-239. There is so much uranium-238 in nature that it could, if converted to plutonium-239, far outstrip fossil fuels as an energy source. Limitless energy supply seemed within the reach of mankind, a prospect that gave rise to fervent, almost religious declamations by scientists about the deliverance of mankind.

The first nuclear engineering achievements were made by the U.S. military's crash program to develop the atom bomb during World War II, known as the Manhattan Project. Uranium-235 provided the explosive energy in the bomb that destroyed Hiroshima; plutonium-239 powered the Nagasaki bomb. The Manhattan Project also showed that it was possible to build large nuclear reactors, to produce plutonium in them, and subsequently to chemically separate the plutonium from fission products and the remaining uranium.

As the United States entered the post-war era, millions of Americans believed that their lives or the lives of soldiers personally near and dear to them had been saved because the atom bombings of Japan had ended the war early. U.S. leaders saw in nuclear weapons the potential to move the world in a political direction of their choosing. The immense technological feats that the U.S. had accomplished during World War II were exemplified most dramatically for all (including Stalin) by the Manhattan Project. Now they would be applied to making the United States by far the most militarily powerful country in human history and also to the material salvation of mankind. Nuclear energy was in the center of that military and economic prospect. America's romance with the atom had begun.[21]

But before commercial nuclear energy could save mankind, some problems, seemingly mundane, remained. They would come to domi-

---

[21] For one description of the military aspects of this "romance with the atom," a phrase coined by Robert Alvarez, see Makhijani et al., eds. 1995, Chapters 1 and 6.

nate the fate of nuclear power. The devil, it turned out, was in the details:

- Nuclear fission gave rise to lighter isotopes of elements, called fission products, that were generally far more radioactive than uranium-235. As a result, the more energy that was generated from a batch of fuel, the more radioactive the fuel became. Severe accidents in nuclear reactors could produce a great deal of harm, and waste disposal from nuclear power would be a serious problem.

- Neutrons, essential to creating chain reactions, also made reactor vessels and other parts intensely radioactive. This could make reactor maintenance and repair as well as disposal of used reactors far more difficult and costly than with conventional power plants.

- The intense radioactivity and high temperatures inside nuclear reactors meant that it would be difficult, complex, and expensive to build, test, and operate them for continuous power production.

- Except in rare ores, uranium was a trace material, and it would take a large amount of ore to produce relatively small amounts of reactor fuel, especially given that uranium-235 was only about 0.7 percent of natural uranium. Further, uranium ore was always mixed with other radioactive materials that would be discharged as waste in large quantities.

- Plutonium produced in nuclear reactors could be used to make nuclear weapons, if separated from spent fuel. Safeguarding plutonium presented challenges that were not encountered with fossil fuels.

- The technological and resource base needed for nuclear weapons was to a very large extent the same as that needed for nuclear power. Thus, the problem of preventing proliferation of weapons while using nuclear power as an energy source would be a crucial security issue.

These problems seem clear enough in hindsight. But how many were apparent in the early days? Was the romance with the atom a case so intense that it blinded engineering judgment? Was it propaganda waged for economic or military purposes? Or was it a mixture of both?

CHAPTER 2:
ELECTRICITY PRODUCTION AND
NUCLEAR REACTORS

An energy source cannot be inexhaustible in the economic sense unless it is priced so low that it can be used in essentially unlimited quantities. After all, solar energy is "inexhaustible" in a physical sense in that we have a continual, huge, and, from a human point of view, essentially endless supply. Yet it is not in widespread use as an energy source because of the relatively high cost of putting it into a usable form, such as electricity. Thus, for solar energy or any other energy source to be "too cheap to meter" it must not only be plentiful in physical terms; it must also satisfy minimal economic criteria. Even fossil fuels resources are huge, if resources such as oil shale are included. But oil shale and similar low-grade resources are generally not included in estimates of the recoverable fossil fuel resource base because they are economically and environmentally unviable.

Let us take a look at the elements of the cost of a large-scale, electricity-generating system, such as would be typical of nuclear power.

Electricity on a large scale is produced by forcibly spinning conducting wires (usually made of copper) through a magnetic field. Such a device is called an electric generator. The energy required to spin the generator and supply the current to the devices that use electricity must come from somewhere. This is the energy source for the electric power station. For instance, falling water is an energy source that is used to spin water turbines, which, in turn, drive electric generators.

The most common energy sources for electricity generation are fossil fuels, which release their energy in the form of heat upon being burned. This heat is converted into mechanical energy in a "heat engine." An internal combustion engine, such as that in a car fueled with gasoline or diesel, is one example of a heat engine. A boiler combined with a steam turbine is another way in which the chemical energy in fuels is converted into mechanical energy.

26

The electricity from a large-scale generating station is transmitted at high voltage (to minimize transmission losses) to the areas where it will be used. Finally, there are extensive networks of wires and transformers that distribute electricity to consumers at the voltages they require for their applications. This scheme is used in all central-station electricity generation.[22]Figure 1 shows the basic elements of a nuclear power plant. The basic arrangement of a coal-fired power plant is the same, except that the reactor and steam-generator are replaced by a coal-fired boiler. (A gas-cooled reactor is shown here.)

Figure 1: Basic Elements of a Nuclear Power Plant with a Separate Steam Generator

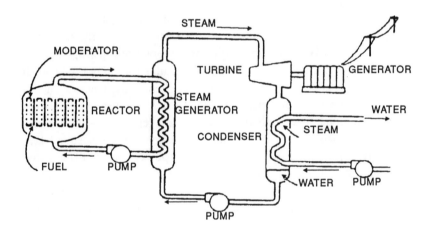

Source: Till and Meyer, eds. 1983, p. 1-21.

The cost elements of an electricity-generation system based mainly on central station plants such as that diagrammed in Figure 1 are:

- capital cost of the power plant, including the boiler and steam turbine (or other source of mechanical energy) to drive the electricity generator and the generation system;

---

[22] Electricity can be generated in other ways, as, for instance, by direct conversion of chemical reations to electriciy in a battery. But such methods are very expensive and so far uneconomical for centralized, large-scale power stations.

- transmission lines;

- distribution network for connecting the main electricity grid of transmission lines to consumers;

- operating and maintenance cost other than fuel;

- fuel cost.

The most important thing to note about this list when evaluating the official claims that nuclear energy could one day be too cheap to meter, is that all the cost elements of a nuclear electricity system other than the fuel would be common between an electric power station that used coal (or another fossil fuel) and one that used nuclear fuel (either uranium or plutonium or some combination of the two).

The principal difference between a nuclear power station and, say, a coal-fired power plant, is in the nature of the fuel. In the one case, it is coal, which is burned in a boiler to generate hot gases, which in turn heat up water to produce steam. The boiler for using coal (or oil or natural gas) is designed to burn the fuel chemically. Nuclear energy does not come from chemical reactions, such as burning, but from nuclear reactions. The nuclear reactor merely replaces the boiler in a conventional fossil fuel power station. It generates the steam that drives the turbine. In other words, a nuclear power station differs from a conventional power station only in the fuel and the details by which the fuel is used in the boiler to generate heat. An important detail here is that the nuclear fuel is much more compact because each fission releases about 200 MeV (megaelectron volts) of energy, while burning one atom of carbon and turning it into carbon dioxide releases about four electron volts (eV). The higher energy per fission means that the volume of nuclear fuel per unit of power output is far smaller than for fossil fuels.

Let us now look at the actual costs of electricity generation at the time that Lewis Strauss made his famous "too cheap to meter" remark. The price of electricity in 1954 to very large industrial consumers (which is close to the cost of generation, since transmission and distribution costs for these consumers tend to be low) was about 1 cent per kilowatt hour of electric energy generated (about 5.7 cents in 1995 dollars using the consumer price index). Subtracting the fuel cost for coal of about 0.4 cents per kilowatt hour (average price of coal, plus average coal transportation cost), we get an estimate of all other aspects of the cost of electricity generation in the mid-1950s other than fuel. This amounts to about 0.6 cents per kilowatt hour in the mid-1950s.

Since all other aspects of electricity generation were common between coal-fired and nuclear power station, the *minimum conceivable charges* for nuclear electricity as calculated for costs prevailing in 1954 would be 0.6 cents per kilowatt hour. Thus, for the largest industrial consumers with factories near generating stations, the costs of nuclear electricity could be expected to be at least 60 percent of the costs of coal under assumptions so optimistic that they were considered unrealistic.

For small consumers, the cost reduction from this most optimistic assessment of nuclear energy would be far lower. This is because transmission and distribution constituted the lion's share of the cost of electricity for households and small businesses, that is, for the overwhelming majority of consumers. The average price of electricity to small consumers in 1954, the year of Strauss's speech, was 2.7 cents per kilowatt hour, of which only about 0.4 cents was the cost of coal (in the case of coal-generated electricity). Thus, even if all fuel costs were eliminated, the average price of electricity to homes and small businesses would still have been 2.3 cents per kilowatt hour or about 85 percent of the full price. That was the best that nuclear energy could be expected to do.

Such cost estimates had, even on the surface, two unrealistic assumptions:

- Nuclear fuel would be so plentiful and so easy to produce that its costs would be insignificant compared to coal.

- Nuclear reactors and associated equipment would cost no more than conventional boilers, despite the greater technical complexity, high energy density, and radioactivity associated with nuclear energy.

Let us take a look at each of these elements of the cost of nuclear power that were readily apparent in the 1950s. (At that time, radioactive waste disposal issues were not forecast to pose serious economic or political constraints on the development of nuclear energy.)

Nuclear Fuel

There are two basic fuels that are used in nuclear power reactors: uranium-235 and plutonium.[23] Natural uranium is the basic raw material for them both. Thorium-232, which occurs in nature, is also potentially a nuclear energy resource. Like uranium-238, thorium-232 is not fissile and cannot sustain a chain reaction. However, neutron absorption by a thorium-232 nuclear converts it into uranium-233 in a manner analogous to the conversion of uranium-238 into plutonium-239. Uranium-233 is fissile and can be used for both nuclear weapons and nuclear power. However, no schemes for using thorium-232 as an energy source have been commercialized. Nor has uranium-233 been used in nuclear weapons, so far as public information indicates.

1. Uranium Fuel

Uranium is ubiquitous in very low concentrations. For instance, it is present in surface waters at concentrations of about 0.7 parts per billion (by weight) and in soil typically at concentrations of two or three parts per million. But it is too costly to extract pure uranium for use in nuclear reactors from such sources. Uranium ores typically contain two-tenths of 1 percent to roughly one-half percent uranium by weight.[24] Therefore, it is necessary to mine 200 to 500 metric tons of ore to get one metric ton of pure uranium. Of this, only about seven kilograms is the fissile isotope uranium-235.

Uranium is present in nature in many different chemical forms. The ores are processed in factories called uranium mills, where the other minerals and materials are separated from uranium. The wastes, containing thorium-230 and radium-226, which are radioactive materials associated with the decay of uranium-238 (see Appendix B), are dis-

---

[23] Two fissile isotopes of plutonium are created in nuclear reactors: plutonium-239, which is the bulk of the plutonium, and plutonium-241. The other two isotopes of plutonium created in reactors in significant quantities are fissionable but not fissile. They are plutonium-240 and plutonium-242.

[24] Uranium ores can have far higher uranium concentration, as much as 10 percent or higher, but they are rare. Ores containing less than 0.2 percent uranium could be mined if uranium prices were higher, as they were in the early 1980s. Residues from refining of gold and copper, which contain as little as 0.01 percent uranium, can be processed to yield uranium as a by-product.

Jackpile Mine, near Grants, New Mexico, is largest open-pit uranium mine in the U. S. Uranium mining has been among the most hazardous steps of nuclear materials production, in terms of radiation doses and numbers of people affected by routine nuclear operations. (U.S. Department of Energy)

charged into tailings ponds. These tailings also contain non-radioactive toxic materials such as arsenic, molybdenum, and vanadium.[25]

Uranium mills produce uranium in the form of uranium oxide (U_3O_8), also called yellowcake.[26] Before it can used in reactors, the uranium must be put into a suitable chemical and physical form and it must have the appropriate content of fissile uranium-235. For most reactors, the proportion of uranium-235 in reactor fuel must be considerably greater than the 0.7 percent concentration found in natural uranium (see below for reactor and fuel types). A large amount of processing is needed to accomplish this. The most expensive step is

[25] The various steps to extract and refine uranium, along with the environmental consequences of each step, are described in Makhijani et al., eds. 1995, Chapter 3.

[26] Pure U_3O_8 is actually a black compound. Yellowcake derives its yellow appearance from the presence of ammonium diuranate in the final product.

uranium enrichment, so called because it increases the proportion of uranium-235 in the fuel. This process produces another stream of uranium, called depleted uranium, which has a uranium-235 content far lower than natural uranium (usually about 0.2 to 0.3 percent uranium-235). Figure 2 shows the steps in converting uranium into a fuel for light water reactors, the most common kind of nuclear reactor used in power generation today.

Figure 2: Converting Uranium into Low Enriched Fuel for Nuclear Reactors

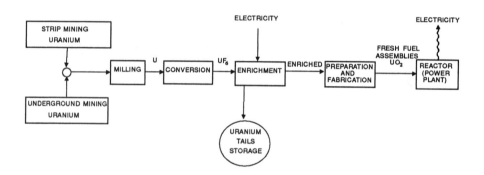

Source: Till and Meyer, eds. 1983, p. 1-20.

As a consequence of the practical necessities of uranium extraction and processing, the reality of the amounts of materials that needed to be handled and processed is far different than the romantic accounts of pellets the size of vitamin pills. While one gram of uranium-235 was equivalent to three metric tons of coal, it typically required 200 grams of natural uranium to obtain a gram of uranium-235 in a practical fuel. And it took on the order of 50 kilograms of uranium ore to produce 200 grams of uranium. Roughly an equal amount of low-grade material littered the mine sites. In sum, about 100 kilograms of ore and rejects had to be unearthed to produced a single gram of uranium-235 fuel. Coal typically came in far richer seams, so that, for high-grade deposits, such as are commonly found in the western United States and elsewhere, the amount of additional material handled at the mine site was not far greater than the end product.

2. Plutonium Fuel

In the minds of its promoters, the promise of endless nuclear energy depended centrally on the conversion of uranium-238 into plutonium-239. A suitably romantic term was given to uranium-238, which was not a fissile material and hence not suitable as a reactor fuel. Uranium-238 was called a "fertile" material because it gave birth to plutonium-239, a fissile nuclear fuel.

As we have noted, uranium-238 is converted into plutonium-239 by bombardment with neutrons. Since a very large number of atoms of uranium-238 nuclei must be so converted to produce substantial quantities of fuel, uranium-238 must be placed in a situation where a correspondingly great numbers of neutrons are being continually generated. This happens in a nuclear reactor when uranium-235 (or another fissile material) is undergoing fission at a suitable rate.

Some of the plutonium produced in a nuclear reactor also undergoes fission, contributing to energy generation. But the rest cannot be directly used as a nuclear fuel because it is mixed with large quantities of unconverted uranium-238, residual uranium-235, and highly radioactive fission products. In order to use plutonium as a reactor fuel (or as a material for nuclear weapons), it must first be separated from the fission products and remaining uranium in the reactor fuel. Table 1 shows an example of one possible composition of reactor fuel when it is inserted into a reactor and the final composition when it is discharged from the reactor (when it is called "spent fuel," although irradiated fuel would be a more accurate term).

Table 1: Composition of Fresh Enriched Uranium Fuel and Spent Fuel

Substance	Initial Percentage by Weight in Fuel	Percentage by Weight in Spent Fuel after 3 Years
Uranium-238	97	95.1
Uranium-235	3	0.8
Plutonium, fissile isotopes	0	0.7
Other plutonium isotopes	0	0.2
Fission products	0	3.2

Adapted from Lamarsh 1993, Fig. 4.25, p. 150. Figures are rounded. Small quantity of uranium-234 present in fresh and spent fuel is not listed because, while it is radiologically important, it is not relevant as an energy source.

The set of steps required to extract plutonium from spent fuel is called "reprocessing" because it involves processing the fuel a second time around (the first time being when the fuel is first fabricated for use in a reactor). Reprocessing is very costly for five reasons:

- Fission products are highly radioactive and must be handled remotely.

- Large quantities of corrosive chemicals are needed to separate the plutonium from the fission products and then from the residual uranium.

- Since uranium and plutonium have similar chemical properties, a large number of steps is required to separate them from each other.

- Since plutonium can be assembled into a critical mass, the processing equipment must be specially designed and all operations carried out with extreme care to prevent accidental criticality.

- Radioactive waste management and disposal is expensive.

A number of plutonium isotopes are created in a nuclear reactor. Once uranium-238 is converted into plutonium-239, some atoms of the latter absorb neutrons and change into heavier isotopes of plutonium, namely, plutonium-240, plutonium-241, and plutonium-242. Plutonium-238 is also created via two different sets of nuclear reactions, one starting with uranium-238 and the other with uranium-235. All these plutonium isotopes, including plutonium-239, are far more radioactive than either uranium-235 or uranium-238. Like natural uranium isotopes, most plutonium isotopes made in nuclear reactors emit alpha radiation, but far more intensely. Alpha radiation consists of fast nuclei of helium, which cannot penetrate the dead layer of the skin. But, when lodged inside the body, alpha particles cause radiation damage to the living cells around them. Plutonium-239 can be relatively easily shielded and is thus hard to detect if it is stolen and removed from the confines of safeguarded facilities. At the same time, it is dangerous to process because small quantities once lodged inside a worker's body greatly increase cancer risk.

The dangers of plutonium were discovered and reasonably well-understood during the course of the Manhattan Project. Their practical

Fuel loading at Peach Bottom atomic reactor. (National Archives)

effect for nuclear power would be that it would be difficult and costly to fashion plutonium into fuel for nuclear reactors due to the protection from radioactivity exposures and the security precautions that would always be needed.

While it was understood that reprocessing would involve substantial costs, the magnitude of these costs was not fully realized until commercial reprocessing was attempted on a large scale from the 1960s onwards and numerous difficulties were encountered in the 1970s. The high cost and unexpected technical difficulties were associated at least partly with the far larger quantities of fission products present in reactor fuel relative to irradiated uranium used for military plutonium production (see below).

At the same time, it was commonly believed until well into the 1970s that uranium was a very scare resource. A corollary belief was that large-scale utilization of nuclear power would necessitate the use of plutonium as a fuel. This view continues to have a large number of adherents in the nuclear establishment despite the high expense of plutonium as a fuel relative to uranium for at least the next few decades.

Nuclear Reactors

> Nuclear power plants, it should be clear, are complex installations and by their nature, they must be designed with care.
>
> —John R. Lamarsh, *Introduction to Nuclear Engineering,* a textbook[27]

As we have discussed, energy from nuclear fission comes from the transformation into energy of a small amount of the mass of a heavy nucleus when it is split. When the nucleus of uranium-235 or plutonium-239 is fissioned, the resulting energy takes many forms. Some of the energy is released in the form of high speed neutrons, some appears as electromagnetic radiation (gamma rays); most is released as vibrational energy of the fission fragments. Almost all this energy is quickly transformed into thermal energy, or heat. A nuclear reactor is basically a vessel that is designed to capture this heat energy in a liquid or gas medium, called a coolant, in a sustained and controlled way. A nuclear reactor must have the following features:

- It must accommodate a sufficient number of fuel rods to sustain a chain reaction at the maximum level of thermal power to be generated. (Power is defined as the rate of energy production).

- It must incorporate ways to control the chain reaction, so that the level of power output can be maintained constant at the required level or varied from zero to the maximum, as necessary, without the danger of severe runaway nuclear reactions.

- There must be ways to capture the energy from the fission reactions and radioactive decay of the fission products and transport it out of the reactor vessel.

---

[27] Lamarsh 1983, p. 119.

- The vessel must be strong enough to withstand high temperatures and (in most cases) high pressures, as well as intense neutron bombardment.

- The vessel and the structure in which it is located must contain the radiation within them so far as possible to minimize radiation doses to workers and off-site populations.

The central function of the nuclear reactor is to generate heat at the required rate in order to drive a heat engine. A number of different reactors have been designed to accomplish this. Another function of reactors is to convert uranium-238 into plutonium-239, although in most commercial reactors this has become a secondary function. In fact, in the context of non-proliferation, it is a problem. Reactors designed specifically to produce more fissile material than they consume as a result of the conversion of uranium-238 into fissile plutonium isotopes are called "breeder reactors."[28]

Reactors are classified into two types: *thermal reactors*, which use thermal (or "slow") neutrons to sustain the chain reaction, and *fast reactors*, which use fast or energetic neutrons to sustain the chain reaction.

1. Thermal Reactors

The design of nuclear reactors depends centrally on the type of coolant that is used to carry off the heat produced in the reactor vessel. For thermal reactors, it also depends on the choice of a material called the *moderator*, which slows down the fast neutrons emitted in the process of fission.

Sustained chain reactions can be achieved with smaller proportions of fissile isotopes in the reactor fuel if the neutrons emitted from fission reactions are slowed down. For instance, some reactors that use slow neutrons can even use natural uranium as a fuel, even though it contains only about 0.7 percent of fissile uranium-235. Slow neutrons, called thermal neutrons, have energies of a fraction of an electron-volt (eV). Neutrons from fission reactions typically have energies of several

---

[28] Reactors that use thorium-232 as the raw material to produce fissile uranium-233 are also possible, but no significant commercial reactors of this type have been built.

megaelectron-volts (MeV) at the time they are emitted.

The process of slowing down neutrons in a nuclear reactor is called *moderation*. It is achieved by putting a moderator in a nuclear reactor. A moderator should preferentially be a light element so that neutrons can slow down when they collide with its atoms. For the most part, this happens by elastic collisions. This process is analogous to that by which billiard balls slow down when they collide with balls of similar weight. Heavy atoms would make less suitable moderators since neutrons would not lose as much energy to them in collisions. This can be visualized as billiard balls simply bouncing off when they collide with the (far heavier) edge of the pool table. Many collisions are needed to slow down fast neutrons to thermal energies. These collisions convert the kinetic energy of the fast neutrons into heat, which is randomized rather than directed kinetic energy. Finally, the moderator must also not absorb too many neutrons in the process of slowing them down. Otherwise sufficient neutrons will not remain to sustain a chain reaction.

Transfer of energy out of the reactor vessel requires that a coolant flow through it. Without a coolant, continued production of fission energy would cause the reactor vessel and its contents to get very hot. This would rapidly lead to a melting of the fuel and fuel rods, a phenomenon called a "meltdown." The coolant must also carry away the heat generated by the radioactive decay of fission products, which build up in the reactor as the fission process continues. When a reactor has been operating for a long time, the heat from decaying fission products alone amounts to several percent of the full power rating. Loss of coolant in a reactor can produce a meltdown in such cases just due to the failure to carry away the decay heat from the fission products. For instance, this was the cause of the partial meltdown in Three Mile Island Unit 2 in 1979.[29]

In some reactors, the coolant and moderator are the same material. Hydrogen is an excellent moderator, being light and having a low neutron absorption cross-section (or probability). However, hydrogen gas is explosive and so it is used in the chemical form of ordinary water, H_2O, also called light water. Further, the density of hydrogen in water (that is, the number of hydrogen atoms per unit volume of water) is far

---

[29] See TMI Commission 1979 for an account of the accident.

Fabrication of light water reactor pressure vessels at the Babcock-Wilcox plant in Indiana. (National Archives)

greater than that of hydrogen gas. Thus, a smaller volume of water gives the same amount of moderation as a far greater volume of hydrogen gas. Besides working well as a moderator, water is also a good coolant. Thus, the most common reactor types in the world use light water as a coolant and moderator. They are called light water reactors or LWRs. Figure 3 (page 41) shows a schematic diagram of one type of light water reactor called a boiling water reactor (BWR). In these reactors, developed by General Electric, the water that serves as a coolant and moderator in the reactor is boiled directly in the reactor. This steam is used to drive a turbine. The main advantage of the BWR design is that it does not require an expensive boiler apart from the reactor. There are a number of disadvantages however, including higher emissions of radioactive gases and the fact that the turbines are exposed to radioactive steam.

Light water reactors are also used in another design, called a pressurized water reactor (PWR). This design, which is the most common power reactor design today, has two water circuits. The primary circuit is the high pressure water in the reactor vessel. This water is kept under such high pressure that it does not boil. The hot, high pressure water is passed though a heat exchanger, called a steam generator, where it heats up water in the secondary circuit and converts it into steam, much as the hot gases in a conventional boiler convert water in a boiler into steam. There are usually three or four steam generators in a PWR. The steam generators add considerable expense to the nuclear reactor but keep the radioactive primary coolant out of the turbines. The line diagram of a nuclear power station in Figure 1 (page 27) shows a power plant with a steam generator. That figure differs from a PWR only in that it indicates a solid moderator, whereas in a PWR the coolant and moderator are the same—ordinary water.

Deuterium, or heavy hydrogen (symbol: D), the nucleus of which consists of one proton and one neutron, can also be used as a moderator. It is the best moderating material from the point of view of low neutron absorption. Like ordinary hydrogen gas, it is explosive and so is used in the chemical form of water, called heavy water (symbol: D_2O). In contrast to LWRs, heavy water-moderated reactors (HWRs) can use natural uranium as fuel. Figure 4 shows a diagram of an HWR used for power generation in Canada, called a CANDU (CANada Deuterium Uranium) reactor.

Carbon in the form of graphite is also a good moderator, but carbon-moderated reactors need a separate coolant. The most common coolants are helium gas, carbon dioxide gas, or water. Reactors of the Chernobyl design (called RBMK reactors) use carbon in the form of graphite as a moderator and water as a coolant.

It is also necessary to control the chain reaction in order to vary the power output of the reactor. To maintain power at a sustained fixed level each fission of a heavy nucleus must produce exactly one more fission. This means that only one of the neutrons arising from fission must give rise to another fission. The ratio of the number of fissions that each fission reaction gives rise to (on average) is called the *multiplication factor*. For a sustained power level, the multiplication factor must be precisely equal to one. At this point, the reactor is *critical* and the nuclear chain reaction will sustain itself at constant power output. If the multiplication factor falls below one, the reactor becomes *subcritical* and the chain reaction will stop. If it rises above one, the reactor is *supercritical* and the power level will increase.

A parameter, called *reactivity*, is often used to describe reactor control. It is related to the multiplication factor in the following way: If the multiplication factor is exactly one, the reactivity is exactly zero; if the multiplication factor is greater than one, the reactivity is positive

Figure 3: A Boiling Water Reactor (BWR)

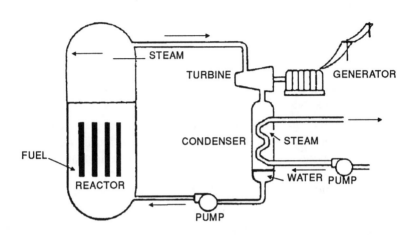

Figure 4: A CANDU Type of Heavy Water Power Reactor

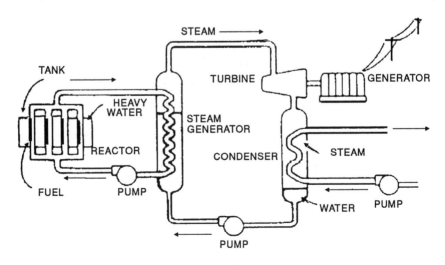

Sources: Till and Meyer, eds. 1983, p. 1-21.

(but less than one). If the multiplication factor is between zero and one, the reactivity is negative. Reactivity is a convenient way to describe reactor control because positive reactivity means a supercritical reactor, zero reactivity means a critical reactor, and negative reactivity means a subcritical reactor.

Start-up, shut down, or change in power level—that is, control—of a reactor is accomplished by changing the reactivity.[30] This is done by controlling the number of nuclear fission reactions per second that typically occur in a reactor. A neutron-absorbing material, like boron, is made into rods ("control rods") which are interspersed with the fuel rods and which can be inserted into or removed from the reactor core.[31] This controls the number of neutrons available for fission reactions and the rate of energy production (or power output). A nuclear reactor can be shut down by making the reactivity negative. This is accomplished by inserting the control rods into the reactor far enough so that they will absorb the quantity of neutrons needed to stop the chain reaction. Raising the control rods temporarily makes the reactivity positive, that is, it makes reactor slightly supercritical for a short period of time, enabling an increase in the power level. The reactor is returned to the critical state (reactivity equal to zero) when the desired level of power is achieved.

Control of a reactor can be lost if the reactor continues to stay supercritical (that is, if the reactivity stays positive) for longer than intended. An increase of the multiplication factor is also called a *reactivity insertion*. The intense heat generated by excess fission could overwhelm the cooling systems, causing a severe accident. The most severe accident in nuclear power history, which occurred in reactor number 4 at the Chernobyl power plant on April 26, 1986, involved a loss of control of the nuclear chain reaction.

The time in which reactor power level increases by a factor of about 2.7 (or more accurately, by a factor equal to e, the base of natural logarithms) is called the *reactor period*. This quantity depends on the design of the reactor and the composition of the fuel. Power reactors

---

[30] See Lamarsh 1983, pp. 280-285.

[31] Reactor control in water-moderated and -cooled reactors can also be accomplished chemically by adding a neutron-absorbing material, generally boric acid, to the water. This kind of control is called *chemical shim*. It is not used by itself, but to supplement the control achieved by use of control rods.

are designed to have long reactor periods in order have slow, smooth increases and decreases in reactor temperature. This minimizes thermal stresses and allows for longer reactor operating lifetime. A typical reactor period in a power reactor would be on the order of one hour.

Control of the reactor is facilitated by the fact that while most (generally more than 99 percent) neutrons from the fission process are emitted essentially at the same time as the fission occurs, a small proportion are emitted after a relatively long time. The former are called prompt neutrons, while the latter are called delayed neutrons. If a reactor becomes critical with only prompt neutrons, the reactor period would be only a tiny fraction of a second, so that control of the reactor would be essentially impossible. But if the reactor is designed so that it does not become critical with prompt neutrons only, then the reactor period and the time available to control it can be increased greatly.

But accidental "prompt criticality" remains a safety concern, since control of the reactor could be lost if a reactor becomes critical with prompt neutrons only. The proportion of delayed neutrons in an LWR is about 0.0065 (that is about two-thirds of 1 percent).[32] So long as the reactivity of the reactor stays below the cents proportion of delayed neutrons, the reactor cannot become prompt critical, and can be controlled. An increase of reactivity above the delayed neutron fraction results in the loss of control of the reactor. For comparison, fast neutron reactors using uranium-233 or plutonium-239 fuel are even more difficult to control, since the delayed neutron fraction is only about 0.0020.

Reactors, such as LWRs in which fuel is loaded in batches, require more complex systems to ensure control because when the fuel is fresh, reactivity increase can be large for a modest movement of control rods. During such periods, reactor control is enhanced by adding neutron-absorbing chemicals to the water. As noted above, this is known as chemical shim.

The ejection of control rods from a reactor that has relatively fresh

---

[32] Lamarsh 1983, p. 286. Reactivity relative to the fraction of delayed neutrons is measured in "dollars" and "cents." One dollar of reactivity occurs when the reactivity is equal to the proportion of delayed neutrons, at which stage the reactor is prompt critical. Evidently, to control the reactor, the reactivity must be kept below one dollar, which is why reactivity for normal reactor operation is measured in cents, with one cent being one-hundredth of the reactivity at prompt criticality.

fuel in it could result in a total loss of reactor control. This is more of a potential problem with batch-fueled reactors, such as LWRs, than with continuous fueled reactors, such as the Canadian heavy water reactor (CANDU).

Commercial light water reactors use uranium fuel enriched to between 3 and 5 percent as a fuel. Graphite or heavy water-moderated reactors can use natural uranium as a fuel. This is a considerable advantage in countries that do not have uranium enrichment plants. It was a principal factor that led a number of countries, including the Soviet Union, France, and Britain, to choose graphite-moderated reactors when they began their military plutonium production. U.S. naval reactors use highly enriched uranium (up to 97.6 percent enrichment) as a fuel because this enables the reactors to operate for longer periods without refueling.

Table 2 (pages 46-47) shows various types of thermal reactors, along with the coolants, moderators, and fuel types they use.

2. Breeder Reactors (Fast Neutron Reactors)

As we have discussed above, of the fissile materials usable for practical nuclear energy production only uranium-235 occurs in any substantial quantities in nature. The other two, plutonium-239 and uranium-233, must be made from uranium-238 and thorium-232 respectively, which are far more abundant than naturally occurring fissile uranium-235. The process of converting "fertile" uranium-238 and thorium-232 into fissile materials is called "breeding," evidently by analogy with biological reproduction.

Commercial nuclear power reactors use natural or "low-enriched" uranium as fuel. Natural uranium contains 0.711 percent uranium-235 and "low-enriched" reactor fuel contains from 1 to 5 percent uranium-235, depending on reactor design. Almost all the rest is uranium-238. (See Appendix B.)

Some of the neutrons in a nuclear reactor convert uranium-238 into plutonium-239. In other words, there is "breeding" of plutonium in all commercial reactors containing uranium-238. However, the term "breeder" reactor is reserved for those reactors in which the production of plutonium-239 (or uranium-233) from fertile materials is greater than the amount of fissile material consumed in the reactor. The ratio of the number of fissile atoms produced to that consumed is called the "breeding ratio" or "conversion ratio." A reactor that is designed so that the breeding ratio can exceed one is called a "breeder reactor." When this happens, the fuel output is greater than the fuel input. This (potential) feature was one of the reasons that nuclear energy was often described as a magical energy source.

In commercial reactors now in operation around the world, like LWRs and HWRs, the breeding ratio is less than one; they are referred to as "converter reactors." Typically, a light water reactor converts just under 2 percent of the uranium-238 into plutonium isotopes, about two-thirds of which consists of the fissile isotopes plutonium-239 and plutonium-241, while the rest consists of the non-fissile isotopes, mainly plutonium-240. Almost half of this plutonium is consumed during normal reactor operation, leaving the rest in the spent fuel. The plutonium consumed during reactor operation typically contributes about one-fourth to one-third of the energy generated in light water reactors.[33]

Theoretically, it is possible to use breeder reactors to vastly increase the amount of fissile material available for future use while producing energy for current use. The amount of time required to double the quantity of fissile material is called the "doubling time." For breeder reactors that convert uranium-238 into plutonium-239, theoretical doubling times are nine to 16 years, depending on reactor design; for reactors that convert thorium-232 into uranium-233, doubling times are estimated at 91 to 112 years. A longer doubling time means that a larger resource base of relatively scarce uranium-235 would be required to create an extensive nuclear energy system.

Since doubling times for breeding U-233 are far longer than for breeding Pu-239, almost all breeder reactors so far have been built to breed Pu-239. A further disadvantage of thorium-232-based breeder reactors cycle is the high gamma radioactivity due to contaminants in recovered uranium-233. This radioactivity arises mainly from the decay products of uranium-232, which is created in thorium-uranium

---

[33] This estimate is calculated as follows: With 3.3 percent enriched uranium fuel, after 30,000 megawatt days of burn-up, the spent fuel contains about 3.3 percent fission products and about 1 percent uranium-235. The energy release per fission for uranium-235 and plutonium-239 is about the same. Since about one out of every 3.3 fissions is plutonium (the rest being uranium-235), about 1/3.3 or 30 percent of the energy comes from plutonium. The fraction of energy from plutonium will vary with fuel enrichment and burn-up. Relative abundance data are from Benedict et al. 1981, Fig. 3.3, p. 88.

Table 2: Basic Characteristics of Reactor Types

Reactor Type	Light Water Reactor (LWR)		Heavy Water Reactor (HWR)
	a) Boiling Water Reactor (BWR)	b) Pressurized Water Reactor (PWR)	
Purpose[1]	electricity	electricity; nuclear-powered ships (U.S.)	electricity; plutonium production
Coolant Type	water (H$_2$O)	water	heavy water (deuterium oxide
Moderator Type	water	water	heavy water
Fuel—Chemical Composition[2]	uranium dioxide (UO$_2$)	uranium dioxide	uranium dioxide or metal
Fuel—Enrichment Level[3]	low-enriched	low-enriched	natural uranium (not enriched)
Comments	steam generated inside the reactor goes directly to the turbine	steam is generated outside the reactor in a secondary heat transfer loop	used in Canada: called "CANDU"—Canadian Deuterium Uranium; was also used in Savannah River Site reactors (metal fuel at SRS)

Source: Lamarsh 1983, pp. 120-143.

Notes:

1. The purpose of the reactor does not depend on the choice of coolant or moderator, but rather on reactor size and on how the reactor operated, and on what ancilliary materials are put into fuel rods besides fuel. The same reactors can, in principle, be used for electricity production, military plutonium production, and production of other radioactive

Reactor Type	Graphite-Moderated Reactor		Fast Breeder Reactor (FBR); Liquid Metal (LMFBR) (most common type of breeder)
	a) Gas Cooled	b) Water Cooled	
Purpose[1]	electricity; plutonium production	electricity; plutonium production	electricity; plutonium production
Coolant Type	gas (carbon dioxide or helium)	water	molten
Moderator Type	graphite	graphite	not required
Fuel— Chemical Composition	uranium dicarbide (UC_2) or uranium metal	uranium dioxide (RBMK) or metal (N-reactor)	plutonium dioxide and uranium dioxide in various arrangements
Fuel— Enrichment Level[3]	slightly enriched, natural uranium	slightly enriched	various mixtures of plutonium-239 and uranium-235
Comments	used in Britain, and France (e.g., AGR, MAGNOX)	used in former Soviet Union, (e.g., Chernobyl [RBMK]; N-reactor at Hanford– now shut)	breeder reactors are designed to produce more fissile material than they consume (e.g., Monju; Phénix)

materials, such as tritium for military and civilian applications. The purposes listed in this column are the common ones to which such reactors are or have been put.

2. Not all fuel types necessarily included.

3. The enrichment of fuel refers to the percentage of the isotope of uranium-235 compared to uranium-238 present in fuel. It is defined here as follows: slightly enriched uranium = about 0.8 to less than 3; low-enriched uranium, 3 to 5 percent.

fueled breeders by various nuclear reactions.[34] India seems to be the only country with a substantial active program to pursue U-233 breeding, since it has very large thorium-232 reserves, which are far greater than its domestic uranium-238 resources.

The number of neutrons per fission required for successful operation of a breeder reactor is considerably greater than for a converter reactor. This is because in addition to the one neutron per fission required to maintain the nuclear chain reaction in the reactor, at least one more is required to convert one atom of U-238 into an atom Pu-239 in order to maintain a breeding ratio of one or more. In practice, since some neutrons are absorbed by the moderator, by other materials in the reactor vessel and by the reactor vessel itself, the number of neutrons required for a breeding ratio greater than one is considerably more than two per fission.

The number of neutrons produced per fission from U-235 or Pu-239 when fissioned by slow (thermal) neutrons is 2.07 and 2.14, respectively; neither of these ratios is sufficiently large to permit the breeding ratio to be greater than one. In other words, there are not enough neutrons available to produce enough plutonium so it will exceed the fissile materials consumed and simultaneously maintain the chain reaction, given other neutron loss mechanisms.

To overcome this problem, breeder reactor designers take advantage of the fact that if the nuclei of U-235 or Pu-239 are bombarded by fast neutrons (energies of several hundred KeV or more), then the number of neutrons per fission increases substantially. For instance, the number of neutrons per fission for 5 MeV neutrons rises to about 3 for U-235 and to about 3.5 for Pu-239. Pu-239 breeder reactors employ this property by using fast neutrons to accomplish both fuel breeding and energy production. Breeder reactors using fast neutrons are also called "fast breeders" or "fast neutron reactors."

Fast breeders, by definition, need no moderators which slow down neutrons, since they use fast neutrons for fission and breeding. They cannot use ordinary water or heavy water as a coolant because these materials also act as moderators. Gases, which have low density, or atoms with heavy nuclei (mass numbers much greater than one), such as sodium metal, can be used as coolants in fast breeders. Molten salt

---

[34] Benedict et al. 1981, p. 368.

has also been proposed. Liquid sodium, which has a mass number of 23, compared to one for ordinary hydrogen and two for deuterium, is the most common breeder reactor coolant. Since a coolant must continually flow across fuel elements, it must be a gas or liquid. Since sodium is a solid at room temperature, it must be maintained in liquid form in a breeder reactor by heating it continually, even when the reactor is shut down.The most common type of breeder reactor is called the Liquid Metal Fast Breeder Reactor (LMFBR). Figure 5 shows a schematic diagram of an LMFBR. A more recent variant of the liquid metal fast reactor design was being developed by Argonne National Laboratory until it was canceled in 1994. It was called the Integral Fast Reactor (IFR). This design had an electrolytic reprocessing plant that accompanied it. Electrolytic reprocessing, called electrometallurgical processing or pyroprocessing, is still being pursued by the DOE at Argonne West in Idaho.[35]

Figure 5: A Schematic Diagram of a Liquid Sodium-Cooled Fast Breeder Reactor

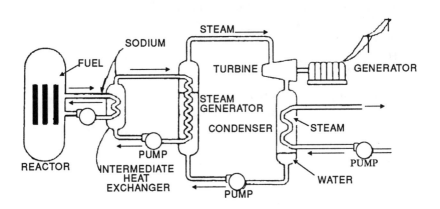

Source: Till and Meyer, eds. 1983, p. 1-21.

---

[35] Sachs 1995, p. 33.

Liquid sodium catches fire on contact with air and explodes on contact with water. Further, the nucleus of ordinary sodium absorbs a neutron and turns into a highly radioactive isotope sodium-24. This is a major threat in case of a breeder reactor accident. To prevent leakage of sodium-24 into the environment, sodium-cooled reactors are designed with two liquid sodium loops. The secondary, non-radioactive sodium loop draws heat from the primary loop and, in turn, is used to boil water in a steam generator. The December 1995 accident at the Japanese breeder reactor at Monju involved a large leak of sodium from the secondary loop.

Despite its theoretical attractiveness in converting non-fissile into fissile material, the breeder reactor has turned out to be a far tougher technology than thermal reactors. Despite five decades of effort during which many pilot and "demonstration" plants have been built, the sodium-cooled breeder reactor design remains on the margin of commercial nuclear technology. The magic of fuel multiplication has not yet been realized on any meaningful scale relative to nuclear electricity-generation levels. Plutonium can also be mixed with uranium for use in thermal reactors. Generally, both plutonium and uranium are mixed after conversion into a dioxide chemical form. For this reason, the plutonium-uranium fuel mixture is called "mixed oxide" fuel, or "MOX" fuel for short.

The "Nuclear Fuel Cycle"

Nuclear power as initially conceived was to be based on using both the natural fissile material uranium-235 and increasing the amount of fissile material by converting uranium-238 (or thorium-232) into fissile materials. In this scheme of things, uranium mining and milling would eventually be a supplement to the creation of fissile materials from an initial stock of fertile uranium-238 and thorium-232 in nuclear reactors.

Reprocessing plants would separate the fissile isotopes from the spent fuel for use in fuel fabrication plants. Many of the long-lived highly radioactive fission products resulting from power generation would be used for a variety of purposes, ranging from nuclear medicine to food irradiation to thermoelectric generators to a vast array of science fiction type of applications that became the subject of much swooning prose in the decade that followed the end of World War II.

There would be little waste. There would be a nuclear fuel cycle.

However, it was recognized even in the early years that large-scale use of nuclear energy would produce fission products in such huge quantities that some arrangements would have to be made for their disposal. But expectations that disposal in salt mines would be a relatively straightforward matter proved too optimisic, like so many other prognostications regarding nuclear power. (See Chapter 6.)

Uranium ore stockpiled at an ore processing plant. Until the mid-1970s, it was thought that uranium was a very scarce resource, a belief which spurred attempts to develop a "plutonium economy." However, uranium has proved more plentiful, and demand for it less strong than anticipated, causing an overall downward movement in prices. (National Archives)

To complicate matters further, reprocessing and fabrication of plutonium into reactor fuel (whether for breeder reactors or light water reactors) turned out to be very expensive, while uranium resources were far more plentiful than anticipated in the 1950s. This made the use of plutonium as a fuel uneconomical, leading to a build-up of spent fuel (which is irradiated fuel discharged from a reactor) at power plant sites. The mounting plutonium stocks, both separated and in spent fuel, are a major source of concern due to their proliferation potential.

CHAPTER 3:
THE EARLY YEARS—ATOMIC MESSIAHS, PROPAGANDISTS, AND SKEPTICS

It is not too much to expect that our children will enjoy in their homes electrical energy too cheap to meter,—will know of great periodic regional famines in the world only as matters of history—will travel effortlessly over the seas and under them and through the air with a minimum of danger and at great speeds,— and will experience a lifespan far longer than ours, as disease yields and man comes to understand what causes him to age. This is the forecast for an age of peace.

> —Lewis Strauss, AEC Chairman, 1954[36]

It should be pointed out that . . . the cost of a nuclear-fuel power plant will be substantially greater than that of a coal-burning plant of similar capacity.

> —AEC report to Congress, 1948[37]

Two strands ran through the promotion of nuclear power in the years immediately after World War II. One was a quasi-religious strain, quite divorced from scientific and engineering considerations. There was another more skeptical side within the nuclear establishment, whose technical assessments tended to be pessimistic or equivocal. After the introduction of hydrogen bombs into the terrible military discourse between the United States and the Soviet Union, the fervent belief in nuclear power was tapped as an instrument of the Cold War.

Atomic Messiahs and Propagandists

The idea that nuclear energy would be extremely cheap and inex-

---

[36] Strauss 1954.
[37] AEC 1948, p. 46.

haustible received a great deal of attention in the immediate aftermath of World War II. For instance, Robert M. Hutchins, the chancellor of the University of Chicago, where the first nuclear chain reaction took place in 1942, said just after the war that an era of cheap and plentiful energy was at hand:

> Heat will be so plentiful that it will even be used to melt snow as it falls. . . . A very few individuals working a few hours a day at very easy tasks in the central atomic power plant will provide all the heat, light, and power required by the community and these utilities will be so cheap that their cost can hardly be reckoned.[38]

From early on in the atomic age, as if in purposeful contrast to the new wartime horrors that could be wrought by the atomic bomb, the future to be brought about by atomic science was depicted in glowing terms to evoke a vision of peace, prosperity, and plenty.

Nuclear chemist Glenn Seaborg, co-discoverer of plutonium and later chairman of the Atomic Energy Commission (AEC), expressed himself with similar enthusiasm. According to Daniel Ford, author of *The Cult of the Atom*:

> The future of civilization, as Seaborg saw it, was in the hands of the nuclear scientists who formed the elite team that would, "build a new world through technology." . . .
>
> Seaborg focused his attention . . . on a visionary dream of atomic-powered plenty. According to his prospectus on its possible applications, nuclear energy was a magician's potion that could free industrial society permanently from all practical bounds. Millions of homes could be heated and lighted by a single large nuclear reactor. . . .The deserts could be made to bloom, sea water could be made potable, rivers diverted—all as a result, he prophesied, of "planetary engineering" made possible by the miraculous new element that he had discovered. . . . There would be nuclear-powered earth-to-moon shuttles, nuclear-powered artificial hearts, plutonium-heated swimsuits for SCUBA divers, and much more. . . . "My only fear [Seaborg stated] is that I may be underestimating the possibilities."[39]

---

[38] Robert M. Hutchins, quoted in Ford 1982, p. 30.

[39] Ford 1982, pp. 23-24.

Lewis Strauss had been a personal assistant to President Herbert Hoover, a Wall Street banker and a personal assistant to Secretary of the Navy James Forrestal during World War II. He rose to become a Rear Admiral. He was appointed chairman of the AEC in 1953.[40] His

Lewis L. Strauss, commissioner, U.S. Atomic Energy Commission. November 1946-April 1950; and chairman, July 1953-June 1958. Strauss coined the phrase "too cheap to meter" in 1954. (National Archives)

---

[40] Hewlett and Holl 1989, pp. 20 and 30-31; and Pringle and Spigelman 1981, p. 115. Hewlett and Holl note that the Senate section of the Joint Committee on Atomic Energy "voted unanimously to recommend Strauss's confirmation without asking him a single question."

passionate advocacy of nuclear power seemed to come from a well-spring of religious sentiment. He had "faith in the atomic future." The progress of nuclear power, he believed, would be guided by "Divine Providence." He believed that the government should create a subsidized "Power Reactor Development Program" because his faith told him "that the Creator did not intend man to evolve through the ages to this stage of civilization only now to devise something that would destroy life on this earth." Later, a member of the congressional Joint Committee on Atomic Energy said of the same period that the government had been trying to "force-feed atomic development" with taxpayer money.[41]

Commissioners of the U.S. Atomic Energy Commission: John G. Palfrey, Dr. Robert E. Wilson, Dr. Glenn T. Seaborg (chairman), Dr. Leland J. Haworth, and Dr. James T. Ramey—November 1, 1962. Seaborg was co-discoverer of plutonium and a strong proponent of nuclear energy. (National Archives)

---

[41] As quoted in Ford 1982, p. 46.

The U.S. Congress caught the fever too. While Democrats like Senator Albert Gore (Vice-President Gore's father) and Senator Clinton Anderson were among the most fervent politicians who were spurring the proponents of nuclear power on the Atomic Energy Commission,[42] by the early 1950s, the "bright promise of the nuclear age had swept over Republicans and Democrats alike in the Congress."[43]

This vision was embodied in the Atomic Energy Act of 1954, the major legislation to define the terms for commercialization of atomic energy in ways that were compatible with the primary military function assigned to it in the context of the Cold War. The Act declares that "the development, use and control of atomic energy shall be directed so as to promote world peace, improve the general welfare, increase the standard of living, and strengthen free competition in private enterprise." Applications of nuclear energy to promote the general welfare were to be "subject at all times to the paramount objective of making the maximum contribution to the common defense and security." [44]

The Act enabled the AEC at its discretion to provide industry with more information. This was a demand of commercial industry and was designed to encourage more industrial participation in the development of commercial applications. In line with the aims of Eisenhower's "atoms for peace" speech, the Act made provisions for international cooperation in commercial nuclear energy development.

Inaccurate and misleading statements, self-righteous justifications, self-delusion, and technological bravado about nuclear power soon became a part of the Cold War hysteria that prevailed in the country. Whatever the reality about the costs of nuclear power, the U.S. wanted to present a benign image of the atom to the world, even as it built a huge arsenal of ever more powerful nuclear weapons. This was a principal motivating force behind President Eisenhower's "Atoms For Peace" program.

1. "Atoms for Peace"

The world's first introduction to atomic energy was the destruction of Hiroshima with an atomic bomb, which instantly reduced much of

---

[42] DOE 1986, pp. 2-8.
[43] Hewlett and Holl 1989, p. 32.
[44] Atomic Energy Act 1954.

that city to rubble and killed about 100,000 people by blast, radiation, and heat.

After the first Soviet nuclear test in 1949, the United States decided to press ahead with the development of the hydrogen bomb. It also opened the Nevada Test Site and began the design and manufacture of nuclear weapons intended for battlefield use. A quarter of a million soldiers took part in atomic maneuvers. The Soviets followed a similar course.

The U.S. tested a thermonuclear device on October 31, 1952, and the Soviets did so on August 12, 1953. People everywhere, including in the United States, were alarmed by the immense potential for destruction of these weapons, which were a thousand times more powerful than the bomb that destroyed Hiroshima. The nuclear arms race had already created the tools that had the potential to destroy the world.

President Eisenhower's December 1953 speech to the United Nations was prepared against this backdrop of U.S. and Soviet nuclear arms development and testing. Initial drafts of Eisenhower's U.N. speech focused on the terribly destructive nature of atomic and thermonuclear weapons. Months before, when Eisenhower took office as president, he was given a report by an advisory group chaired by Robert Oppenheimer, the scientific leader of the Manhattan Project, urging the initiation of "Operation Candor" to convey to the people of the United States and the world in a frank manner the immense dangers and utter destruction the world faced if thermonuclear weapons were used. The report compared the nuclear-armed U.S. and the Soviet Union to "two scorpions in a bottle, each capable of killing the other, but only at the risk of his own life."[45]

Eisenhower's speechwriter, C.D. Jackson, produced a version of the speech which followed the metaphor of the two scorpions in a bottle, calling it the "Bang! Bang!" papers. But Eisenhower was not satisfied with a message which "leaves everybody dead on both sides." Rather than "scare the country to death," he preferred a message that would "find some hope."[46] As a result the speech was revised. One part contained graphic descriptions of the power and terror of nuclear weapons; another part spoke in glowing terms about the promise of the

---

[45] Williams and Cantelon, eds. 1984, p. 73.

[46] Pringle and Spigelman 1981, p. 121; and Williams and Cantelon, eds. 1984, p. 73.

peaceful atom.

Thomas Murray, an AEC commissioner, saw clear propaganda benefits in diverting attention from bombs to civilian power since both the U.S. and the Soviet Union were rushing headlong into the era of thermonuclear weapons. In September 1953, soon after the Soviet hydrogen bomb test, Murray wrote to AEC chairman Strauss, stating that the United States could derive "propaganda capital" from a highly publicized announcement regarding the recent U.S. decision to develop a civilian atomic power plant. He argued that the Soviet hydrogen bomb test offered a "providential" chance for a "great propaganda effect" and urged the U.S. Government to capitalize on this chance by making public its decision to build the civilian reactor "with appropriate fanfare."[47] The propaganda would show the United States as the promoter of the peaceful uses of nuclear energy in contrast to the horror of the Soviet thermonuclear program.

In addition to the propaganda advantage gained by moving into civilian atomic energy and casting the Soviets as the militaristic side (despite the parallel development of Soviet nuclear power plants), another aspect of U.S. urgency to embark on large-scale civilian nuclear energy generation was the fear that, if the U.S. delayed, the Soviets would be the first to achieve it. (As it turned out, the British did it ahead of the U.S., in 1956.) The chairman of the congressional Joint Committee on Atomic Energy, Sterling Cole, expressed this fear in melodramatic tones:

> It is possible that the relations of the United States with every other country in the world could be seriously damaged if Russia were to build an atomic power plant for peacetime use ahead of us. The possibility that Russia might actually demonstrate her "peaceful" intentions in the field of atomic energy while we are still concentrating on atomic weapons could be a major blow to our position in the world. It could even disrupt the continued operation of our own weapon plants by stimulating friendly countries to cut off vital uranium they now sell us—cut it off to avoid the charge at home that they are selling their atomic power birthright for American dollars.[48]

---

[47] Murray 1953b.
[48] Cole 1953.

Such considerations led Eisenhower to focus a large part of his speech, delivered to the United Nations General Assembly on December 8, 1953, on the practical possibilities of civilian nuclear power development. It became known as the "Atoms for Peace" program:

> The U.S. would seek more than the mere reduction or elimination of atomic materials for military purposes.
>
> It is not enough to take this weapon out of the hands of soldiers. It must be put into the hands of those who will know how to strip its military casing and adapt it to the arts of peace. . . .
>
> A special purpose would be to provide abundant electrical energy in the power-starved areas of the world. Thus the contributing powers would be dedicating some of their strength to serve the needs rather than the fears of mankind.[49]

Countries would contribute fissionable materials to a new international atomic energy agency to be created under the auspices of the United Nations. This agency would prevent proliferation of nuclear weapons, and at the same time assist in the development of nuclear power:

> The United States knows that if the fearful trend of atomic military build-up can be reversed, this greatest of destructive forces can be developed into a great boon for the development of mankind.
>
> The United States knows that peaceful power from atomic energy is no dream of the future. That capability, already proved, is here now—today. Who can doubt, if the entire body of the world's scientists and engineers had adequate amounts of fissionable materials with which to test and develop their ideas, that this capability would rapidly be transformed into universal, efficient and economic usage.[50]

Eisenhower also outlined the functions of the new agency in allocating fissionable material and in providing experts around the world

---

[49] Excerpt from speech reprinted in Williams and Cantelon, eds. 1984, pp. 109-111.

[50] Eisenhower's Atoms for Peace speech, as reproduced in Williams and Cantelon, eds. 1984, p. 110.

for development of all technological fields. He finished by promising that the United States would "devote its entire heart and mind to find the way by which the miraculous inventiveness of man shall not be dedicated to his death, but consecrated to his life."[51]

Eisenhower's statement that nuclear power could "rapidly be transformed" from a developmental technology into a "universal, efficient and economic usage" was not based on sound analysis. Rather, it converted the early messianic statements about nuclear power into a calculated tool in the Cold War. His claim about economical nuclear power that would be universally available was widely promoted in what Pringle and Spigelman have described as a "frenzy of public relations activity following the speech," which was widely distributed in ten languages. Further, " . . . 350 U.S. foreign language newspapers and related ethnic organizations started a campaign to ensure that excerpts were sent by immigrants to their relatives and friends abroad." As a result of the campaign:

> Newspapers and radio and TV stations were deluged with briefings, magazine articles, and special advertising. The United States Information Agency distributed the speech and an endless stream of follow-up feature stories. Voice of America radio recordings were given to newspapers and radio stations throughout the world. The sphere of the peaceful atom expanded rapidly as the American broadcasts produced talks entitled "Nuclear Device in Fight Against Cancer," "Forestry Expert Predicts Atomic Rays Will Cut Lumber Instead of Saws," and "Atomic Locomotive Designed." A range of new promotional films was conceived: *Atomic Greenhouse, Atomic Zoo, Atom for the Doctor*.[52]

Lewis Strauss's "too cheap to meter" remark in 1954 was made in the context of the propaganda effort that followed Eisenhower's speech. But there was more than talk. By November 1954, a contract with a Pennsylvania utility, Duquesne Light Company, was in place to build the first power reactor to be installed by an electric utility.[53] Cautions that the development of nuclear power should proceed more

---

[51] Eisenhower's Atoms for Peace speech, as reproduced in Williams and Cantelon, eds. 1984, p. 111.

[52] Pringle and Spigelman 1981, pp. 123-124.

[53] Duncan 1990, p. 198.

deliberately with due attention to the long-term aspects were cast aside (see Chapter 4).

Atomic Skeptics

Anecdotal evidence indicates that many people still believe that there was some scientific and engineering basis for the official pronouncements of the 1940s and 1950s about the potential of nuclear energy to inaugurate an era of unprecedented plenty. However, a review of technical documents produced by economists, engineers, and physicists during the 1940s and 1950s indicates that belief in a future of nuclear energy that would be cheap, let alone far cheaper than coal, was not evident in either governmental or corporate assessments. In fact, we have not found a single serious assessment (as distinct from statements and speeches) from the period that concluded that nuclear power would one day be "too cheap to meter" or even merely cheap.

Before we examine the evidence in governmental, industry, and academic studies of the time, it should be noted that there appears to have been a deliberate element of deception in the statements about nuclear power that even predated the "Atoms for Peace" campaign. A December 1950 speech to the American Association for the Advancement of Science, reprinted in an industry journal, *Nucleonics*, by C.G. Suits, Vice-President and Director of Research of General Electric, is worth quoting a length in this regard, not least because G.E. was one of the two companies that took considerable losses to establish light water reactor technology:

> In addition to laboratory work on power-producing reactors, there has been a vast accompaniment of public debate, discussion and comment on this subject. Perhaps never before has so much been said by so many with so few facts. . . .

> It is safe to say . . . that atomic power is *not* the means by which man will for the first time emancipate himself economically, whatever that may mean; or forever throw off his mantle of toil, whatever that may mean. Loud guffaws could be heard from some of the laboratories working on this problem if anyone should in an unfortunate moment refer to the atom as the means for throwing off man's mantle of toil. It is certainly not that!

.   .   .

. . . . At present, atomic power presents an exceptionally costly and inconvenient means of obtaining energy which can be extracted more economically from conventional fuels . . . The economics of atomic power are not attractive at present, nor are they likely to be for a long time in the future. This is expensive power, not cheap power as the public has been led to believe.[54]

Such frank talk about the economic prospects of nuclear power, which for Suits was primarily a technology with military objectives that may have long-term application to commercial power, seems to have been reserved for those who had been initiated into the world of science and technology. The public at large had little clue.

Initial assessments of nuclear power, such as the one done within the Manhattan Project in 1944, tended to be pessimistic, because it was thought that uranium was an extremely scarce resource.[55] Many public and private, as well as joint governmental and corporate efforts to evaluate nuclear power, were created after World War II. These studies were accompanied by research on and development of various power reactor designs. While reactors had been built during the war for the purpose of producing plutonium for weapons, they were not designed to produce electrical power. Rather, the heat generated by the fission reactions in them was simply dissipated into the environment.

1. Early Practical Assessments

One early detailed public study of the economics of nuclear power, published in 1950, estimated a "*lowest conceivable* cost" (assuming zero cost for nuclear fuel, as discussed in Chapter 2 above) for nuclear-generated electricity at 0.4 cents per kilowatt hour (in 1946 dollars). The same source estimated conventional coal costs at 0.5 to 0.75 cents per kilowatt hour, including fuel cost.[56] Thus, it was already clear that even apart from any considerations of the complexity of building nuclear power plants, even if nuclear fuel were completely free, the electricity generated would cost at least 50 to 80 percent of the cost of conventional coal-fired electricity, exclusive of transmission and dis-

---

[54] Suits 1951.

[55] Fermi et al. 1944. This study was also called the Compton Report.

[56] Schurr and Marschak 1950, pp. 24-25, and 64. Schurr and Marschak assumed a 75-to-100 MWe plant operating at 50 percent capacity.

tribution costs.

Another major cost consideration for nuclear power plants, in addition to operating, maintenance, and fuel costs, was possible breakdowns in the complex new parts and technology. Another early assessment, prepared by Dr. C.H. Thomas of Monsanto Chemical Company in late 1946, put the cost of building a 75 megawatt nuclear power plant at $25 million, or $333 per kilowatt. He reckoned that such a plant could generate electricity for 8 to 10 mills (0.8 to 1.0 cents) per kilowatt hour.[57] The cost of electricity from a coal-fired plant was estimated at 7.5 mills per kilowatt hour.

A number of experts from that time felt that these projections might be too optimistic. In 1947, M.C. Leverett from Clinton Laboratories (later renamed Oak Ridge National Laboratory) stressed the higher labor costs for nuclear power, and indicated that Thomas's estimate might be low:

> This opinion is based on the observation that even when the details of all processes in a plant are worked out, cost estimates made by the best estimators are often seriously low. In this case the details are not worked out, and . . . necessarily based upon certain assumptions which are not complete certainties.[58]

After the initial burst of optimism, skepticism grew regarding low or even moderate cost figures, due to increasing recognition of the technical difficulties which nuclear power faced.

As early as 1947, the chairman of the Atomic Energy Commission, David Lilienthal, warned that it was not appropriate to look to nuclear energy to solve electricity problems in the near future. He was not willing to give a date when nuclear power might be commercial, and would only guess that a demonstration plant of unknown economics might be in operation in "eight to ten years."[59] This was, he said, because of the "jungle of difficult scientific and engineering problems" that would have to be solved.[60] And even then, there would be considerable barriers to widespread use of nuclear energy, notably due to the subordination of civilian to military applications and the extensive

---

[57] Sporn 1950; and Goodman 1949.
[58] Leverett 1947.
[59] Lilienthal 1947, p. 37.
[60] Lilienthal 1947, p. 34.

secrecy surrounding military work. The primary research and development of nuclear technology at the time was taking place within the military framework.

A leading advisor to the AEC on nuclear power, Dr. Lawrence Hafstad, who became director of its Division of Reactor Development in 1949, asserted that "'all of the big' reactors 'run into lots of money.'" Dr. Hafstad seemed to think that the rough formula then in vogue around the Commission—"a megabuck per megawatt"—was about right.[61] At a million dollars per megawatt, the capital costs of nuclear power would be at least seven times those of coal. This would far outweigh any possible advantage that nuclear power might have in low fuel costs.

In 1948, the AEC presented a report to Congress in which it cited "unwarranted optimism as to the character of the technical difficulties [facing nuclear power] and the time required to surmount these difficulties."[62] This report was prepared by an AEC General Advisory Committee, the members of which included many of the leading nuclear scientists and engineers, including Enrico Fermi, Glenn Seaborg, and J.R. Oppenheimer. This committee was not even uniformly optimistic about fuel costs, the greatest economic advantage of nuclear power. It concluded that nuclear reactors would operate at a lower cost than coal, "at least as far as fuel expenditure is concerned," only "if favorable assumptions are made about the cost of uranium and the technical practicality of breeding."[63] In any case, the committee's report warned that "the cost of a nuclear-fuel power plant will be substantially greater than that of a coal-burning plant of similar capacity."[64] It is fascinating that Seaborg, who expressed messianic ideas about the potential of nuclear energy, was also on the panel that was pessimistic about the costs of nuclear power. The public pronouncements were strangely at odds with the hard-headed assessments.

Philip Sporn, the chairman of the AEC's Advisory Committee on Cooperation between the Electric Power Industry and the AEC, noted the basic issues succinctly and clearly in 1950:

---

[61] Lilienthal 1947, p. 34.

[62] AEC 1948, p. 43.

[63] AEC 1948, p. 46.

[64] AEC 1948, p. 46.

If the cost of generating power were dependent only on the price of fuel, here would be a Utopia, indeed. But the processing of the [nuclear] fuel will involve elaborate and therefore relatively expensive chemical operations. Thus, the capital costs of the chemical plant, the nuclear reactor, and the steam and electrical portions of the complete atomic power plant are bound to be the dominant portions of the cost of electrical energy at the point of generation.[65]

Opinion in industry was about the same as that in government. In 1950, Ward Davidson, a research engineer for Consolidated Edison Company of New York, one of the biggest electric utilities in the United States, published a paper in the journal *Atomics*, updating a 1947 opinion about nuclear power. He stated in 1950 that the technical problems facing nuclear power were even more difficult than he had imagined. He specifically pointed out the stringent requirements for materials to be used in a nuclear reactor. They would have to withstand a high temperature environment, to have high tensile and compressive strength, to be workable enough to be manufactured into reactor parts and vessels, to have low neutron absorption (so as not to stop the chain reaction), and to survive the intense neutron bombardment without serious damage. "None of the commonly used engineering materials meet these requirements."[66] He also noted that it would be difficult to manufacture the alloys which might be employed to the degree of quality assurance that was likely to be needed. Small variations in composition may produce unacceptable changes in properties. Moreover, the durability of materials under intense neutron bombardment had hardly begun to be investigated "because adequate testing facilities have not been in existence."[67]

Many of the technical difficulties arose from the constraints on reactor design posed by considerations of power density, which is the amount of heat generated by fission reactions and radioactive decay heat of fission products per unit volume of the reactor vessel. Low power density would make it easier to build a reactor, because the heat generation per unit volume would be low. But it would also increase the size of the reactor vessel per unit of power, and hence tend to make

---

[65] Sporn 1950, p. 305.
[66] Davidson 1950, p. 321.
[67] Davidson 1950, p. 321.

it more expensive. Increasing power density would reduce reactor vessel volume, but would increase the cost of materials and reactor construction due to kinds of problems enumerated by Ward Davidson. What reactor design might be appropriate under the circumstances was still very much an undecided question. There was simply not enough research to answer the crucial questions.

There were other serious questions, not faced in fossil fuel power plants, that remained to be answered. These included control of the reactor (to prevent accidents) and shielding repair and maintenance workers from intense radiation in certain reactor areas.

Davidson's was only one of a number of sobering evaluations of the prospects of nuclear power. Most serious governmental, industrial, and academic assessments done in the first decade after World War II concluded that nuclear power would be uneconomical for a considerable period unless it was subsidized in some way. Of particular interest was the idea that subsidies could be provided by using "dual-purpose reactors" to produce plutonium for military purposes at the same time as they produced power for the civilian economy.

One prominent study was undertaken by four industry-utility teams. They included Bechtel, Monsanto, Dow Chemical, Pacific Gas and Electric, Detroit Edison, and Commonwealth Edison. The study, completed in 1953, concluded that nuclear power would not be economical by itself in the "very near future" and that government purchases of plutonium for military purposes would help:

> All four groups concur in the belief that dual-purpose reactors are technically feasible and could be operated in such a fashion that the plutonium credit would reduce the cost of power. Conversely, all agree that no reactor could be constructed in the very near future which would be economic on the basis of power generation alone.[68]

Significantly, the study did not find that fuel could be considered a very small cost of nuclear power, unless this cost was offset by plutonium production for the military. The Pacific Gas and Electric-Bechtel team estimated that fuel costs would range from 0.13 cents to 0.45 cents per kilowatt hour, depending on reactor type.[69] At the

---

[68] *Nucleonics* 1953, p. 49.

[69] *Nucleonics* 1953, p. 59.

upper end of that range, nuclear fuel costs would be higher than the corresponding costs for coal-fired power generation, which were about 0.4 cents per kilowatt hour in the mid-1950s. The Dow Chemical-Detroit Edison team concluded that "[a]t the present time nuclear power is handicapped by high capital costs, high processing costs, and significant fuel costs."[70] Thus, the most important supposed advantage of nuclear power, a very compact fuel that was popularly presumed would cost very little or next to nothing, was in substantial doubt even as fantastic public pronouncements were being made about the promise of nuclear energy.

Chauncey Starr, one of the foremost proponents of nuclear power, calculated that, for civilian breeder reactors (one of the kinds commonly considered as a promising commercial reactor type in those days), it was "not unreasonable to assume that improvements in the cost of liquid-metal equipment and reactor construction can be achieved." Even so, he observed, "it is not likely that significant reduction in operating costs of the nuclear plant will occur." He further stated that "based upon what we consider a realistic analysis of the future possibilities of atomic energy, we can produce electrical power at a cost almost competitive with conventional present-day plants." He did speculate that future increases in the cost of conventional fuels would probably create an economic role for nuclear power.[71]

As another example, a 1953 AEC study, "Reports to the U.S. Atomic Energy Commission on Nuclear Power Reactor Technology," was unequivocal regarding the short-term prospects for nuclear power: "No reactor could be constructed in the very near future which would be economic on the basis of power generation alone."[72] These words were identical to those in the industry study.

In 1953, the AEC contracted with the Massachusetts Institute of Technology to assess the costs of nuclear electricity without the subsidy of military plutonium sales. According to Mullenbach, an AEC economist, the study, called Project Dynamo, "held out the hope that nuclear power, after long development, might become competitive 'without the crutch of weapon-grade plutonium sales.'"[73] Mullenbach

---

[70] *Nucleonics* 1953, p. 60.

[71] Starr 1973.

[72] As quoted in Mullenbach 1963, p. 55.

[73] Mullenbach 1963, p. 59.

further noted that the "project's results suggested that a substantial program of research, development and small experimental plant investigations would be necessary to avoid the serious risks of premature, full-scale plants."[74]

Although many were skeptical that nuclear reactors could be made economically viable on their own, there was enthusiasm for developing multi-purpose reactors which would allow power generation to piggyback on military uses for reactors: plutonium production and naval ship propulsion.

The negative assessments were based on more than paper studies. They derived from wartime experience with building nuclear reactors and from a number of research and development efforts undertaken right after World War II, such as the mercury-cooled, plutonium-fueled breeder built at Los Alamos in 1946, the experimental breeder reactor built at Argonne West, Idaho, and the development of thermal reactors for naval propulsion.

The heavy involvement of government-owned laboratories in the early years of nuclear power was, in large measure, due to the secrecy imposed on all matters related to nuclear energy in the immediate post war period. A frustrated David Lilienthal, who had just resigned as AEC chairman wrote in 1950 that "no Soviet industrial monopoly is more completely owned by the state than is the industrial atom in free-enterprise America."[75] Lilienthal's article intensified a campaign to open up atomic energy to private enterprise, and led to the 1954 revision (cited above) of the Atomic Energy Act of 1946 that had mandated that secrecy. But it would be the military that would make the vital decisions about the future of nuclear power in the United States, as it did in other countries.

---

[74] Mullenbach 1963, p. 59.

[75] Lilienthal, as quoted in Mazuzan and Walker 1984, p. 18.

CHAPTER 4:
PLUTONIUM, THE NUCLEAR NAVY, AND NUCLEAR POWER DEVELOPMENT

By the mid-1950s, both industry and government had concluded that nuclear would not be economical in the short run, and that in the long run it would perhaps turn out to be competitive with coal. Moreover, coal was plentiful. There appeared to be no urgency to develop nuclear power on that account. But the money was still attractive, if the government would come up with it. And it was a useful propaganda tool in the Cold War. Here is the rationale provided by *Nucleonics*, a nuclear industry journal:

> All are in essential agreement that there is an urgency for this country to develop a strong civilian nuclear power development program. The urgency for this comes not so much from the need for a new, low-cost source of energy as from the vital necessity for this country to maintain international preeminence in reactor technology.[76]

There were two other reasons for government support for nuclear reactors. First, the United States decided after the first Soviet nuclear test to greatly expand its nuclear weapons production capability. Second, the Navy was discovering the potential strategic benefits of nuclear propulsion. This was the basis for the collaboration of government and industry. Private companies had run nuclear weapons plants even during the Cold War and would continue to do so for decades. They were also at the center of the Navy's reactor development program.

---

[76] *Nucleonics* 1953a.

Round One: Dual-Purpose Reactors

The proposals of the 1990s for a DOE reactor that would serve the dual purpose of nuclear weapons materials production and civilian power production have precedents in the 1950s. In Chapter 3, we discussed a number of studies that concluded that nuclear power could only be economical in the short term if it were subsidized by the military, which would also use the plants to make plutonium for nuclear bombs. Some initial industry plans for nuclear power were based on this premise. For instance, according to Hewlett and Anderson, in the spring of 1950, Charles A. Thomas of the Monsanto Chemical Company proposed "that industry be allowed to design, construct and operate atomic power plants at its own expense, to produce both useful power and plutonium."[77]

But the military establishment was in a rush. The 1949 Soviet nuclear test and the Korean War had led to a decision to hugely expand the U.S. nuclear arsenal. It was much easier and faster to build nuclear reactors for military plutonium production only, especially since cost was not a prime consideration. Further, the reactors would be kept entirely within government ownership. Second, those who wanted atomic secrets to be unlocked for civilian purposes would then pursue their goals outside the nuclear weapons production system and not slow down military production. Third, dual-purpose reactors were more difficult and expensive to build, since nuclear reactors that generate electrical power must be operated at much higher temperatures than reactors which are used only to convert uranium-238 into plutonium-239. Higher temperatures mean a more elaborate and expensive reactor that would take longer to design and build. That was not compatible with the military's nuclear aims of increasing production rapidly.

These factors led the AEC to opt for its own dedicated, single-purpose military plutonium production reactors, which it built at the wartime Hanford site in Washington state and the new Savannah River Plant (now renamed the Savannah River Site) in South Carolina. Construction on these began at various dates in the late 1940s and early 1950s and was complete by the mid-1950s. In 1953, the AEC issued its Nuclear Power Policy Statement abandoning the dual purpose reac-

---

[77] Hewlett and Duncan 1990, p. 437.

tor as the basis for developing nuclear energy in the U.S. "It is the objective of this policy to further the development of nuclear plants which are economically independent of Government commitments to purchase weapons-grade plutonium."[78]

The AEC decision to separate military plutonium production from civilian power development was a blow to the section of the nuclear power industry that believed that subsidies from military plutonium production were essential for the development of nuclear power. According to AEC economist Mullenbach, this AEC policy was "an important reason that certain companies, such as Monsanto, lost interest in the program."[79] In 1950, Monsanto had proposed to the AEC that the industry build dual-purpose reactors.

One dual-purpose reactor was built in the United States: the N-reactor at Hanford. It is a graphite-moderated, water-cooled reactor, design features it shares with the ill-fated Chernobyl reactor. A large reactor, with a capacity of 4,000 megawatts thermal, it began operation at the end of 1963. In 1966, it began supplying steam for electricity generation to the 860 MWe electrical generating station built on the Hanford site by the Washington Public Power Supply System.[80] The Chernobyl accident prompted a safety review. The N-reactor was shut down in 1987 and put on cold standby in 1988. It is not expected to operate again.

Despite the fact that the N-reactor was a dual-purpose reactor, a feature that many in the industry felt was needed to make nuclear power economical, its graphite-moderator design was not the one used in the U.S. nuclear power program. The light water reactor was already preferred by the U.S. Navy as the route for civilian reactor development in the United States. The principal route to nuclear power subsidies in the United States, unlike several other countries, would not be via military plutonium production.[81]

---

[78] AEC 1953, p. 19.

[79] Mullenbach 1963, p. 58.

[80] Cochran et al. 1987a, pp. 19-21.

[81] Many dual-purpose reactors were built in other countries, such as the Soviet Union, France, and Britain. (See Makhijani et al., eds. 1995.) Their existence in Russia has created some serious and expensive problems after the end of the Cold War. The United States would like Russia to shut down all military reactors, but since they also supply heat and electricity, and alternative sources are not readily available, they

Round Two: Admiral Rickover, the Nuclear Navy, and the Light Water Reactor

Soon after World War II, the Navy initiated a program to develop nuclear propulsion for surface ships and submarines. Nuclear reactors were attractive to the Navy because they could eliminate the need for refueling stops abroad. They possessed an added advantage for submarines in that nuclear submarines could stay deeply submerged for months at a time, making detection very difficult. (Engines operating on oil need oxygen to burn it, which they get from the air. While submerged deeply they operate on batteries which severely limits the time that they can stay deep underwater. Unlike fossil fuels, nuclear fission does not need air to release its energy.)

From the start of the Navy nuclear propulsion program in 1946, its technical leader was Captain (later Admiral) Rickover. Through that connection he came to dominate the decision regarding the design of the first civilian reactor. The design he chose early on constitutes today the most common type of civilian power reactor in the world.

The Navy had been working closely with Westinghouse in order to develop reactors for ships. The design that Westinghouse was developing involved a pressurized water reactor (PWR), which used light (that is, ordinary) water as a moderator. As we have discussed in Chapter 2, the PWR is a variant of a basic reactor design that uses light water both as moderator and coolant. The PWR was to be used in surface ships as well as submarines.

1. Propaganda Aspects

By late 1953, there was intense pressure to build a nuclear power reactor quickly in order to show that the U.S. nuclear program was a symbol of peace while that of the Soviet Union was militaristic. Westinghouse, which had a major investment in the pressurized water reactor design for the Navy, also wanted to see that design adopted for the potentially much larger civilian nuclear power market.

The perceived urgency of building a large-scale plant for propa-

---

continue to operate. The United States is to provide considerable financial assistance to enable Russia to secure replacement energy sources so that the dual-purpose reactors can be shut by the year 2000. This program is unlikely to accomplish that goal.

ganda reasons was given more weight than the economics of the reactor or long-term considerations about the suitability of the design. The AEC made the decision to proceed with Westinghouse's PWR for the first large-scale civilian reactor in the spring of 1953, despite consid-

Admiral Hyman G. Rickover on board the *USS Nautilus*, 1958. Rickover was the technical leader of the Navy's nuclear propulsion program, and from 1953 to 1957 oversaw construction of the first large-scale civilian power reactor. The decisions he made played a large role in the development of pressurized water reactors as the primary design for reactors in the U.S. (National Archives)

erable opposition. As summarized in the official history of the plant,

> Opposition to Rickover and pressurized-water technology, how-
> ever, was intense. Within the division of reactor development
> the argument ran that the technical approach, while suitable for
> naval propulsion, was too wasteful of uranium-235 for its use
> to be encouraged in civilian applications. By sponsoring a pres-
> surized-water plant to gain international prestige, the commis-
> sion risked distorting the development of nuclear technology
> . . . .[82]

H.C. Ott of the AEC's division of reactor development strenuously
objected to choosing one reactor design prematurely, saying that in
making the assumption that the PWR was to be the basis for going
ahead, there had been "no attempt to justify this project as a logical
part of the over-all reactor development program and no arguments are
advanced to support the thesis that a prototype power plant should be
built."[83]

An AEC classified report given to the Joint Committee on Atomic
Energy stated that the PWR was not advisable from the standpoint of
economics. According to the official DOE history of the period:

> The classified report of more than 130 pages which the Com-
> mission delivered to the Joint Committee in February 1954 . . .
> included reasonably candid evaluations of the status of each
> [reactor design] concept. The pressurized water reactor seemed
> most likely to be successful in the short term, by the end of
> 1957, but it offered a poor long-term prospect of producing eco-
> nomic nuclear power.[84]

Since Cold War propaganda was a principal factor in the decision
to build a commercial nuclear power plant, short-term viability gave
the PWR an edge over other less-developed designs. PWR technology
also had the advantage, in the short- and medium-term, of giving the
U.S. more control over nuclear power development abroad by limiting
fuel availability. Light water reactors require enriched uranium, while
reactors moderated by heavy water or graphite can be run with natural
uranium. Since the U.S. was at the time the only capitalist country with

---

[82] Duncan 1990, p. 197.

[83] Ott 1953.

[84] Hewlett and Holl 1989, p. 195.

uranium enrichment capacity, the reasoning was that developing a technology requiring enriched uranium fuel would better enable the U.S. to maintain control over fissile materials and enforce safeguards against their diversion for use in weapons.[85]

Public relations aspects were important throughout the commercial project. For example, in May 1955, there was a proposal to accelerate the project, because it was feared that the British might complete their reactors earlier and, presumably, get the propaganda advantage. As Lewis Strauss wrote at the time:

> I mentioned to Admiral Rickover my apprehension that the Calder Hall plants of the British Atomic Energy Authority might come "on stream" before the completion of Shippingport and asked him what steps could be taken to accelerate the date of completion of Shippingport.[86]

There was also a pork-barrel aspect in the timing of the decision to build a PWR. The civilian project was based on a Westinghouse nuclear reactor project for an aircraft carrier begun in 1952, which had been canceled by the Pentagon in May 1953. The civilian PWR, in effect, replaced the canceled naval reactor.

In sum, political factors having to do with Cold War and pork-barrel considerations made long-term reactor development a secondary consideration. The internal opposition to the decision to proceed with the PWR was overruled.

2. Shippingport

In June 1953, the AEC assigned the first civilian reactor project to Rickover. The first commercial power reactor in the United States would build a PWR. In October of that year, the Navy contracted with Westinghouse's Bettis lab to design and build the reactor. Rickover also sought a utility to be partner in the enterprise. After screening nine

---

[85] Hewlett and Holl 1989, p. 198. For instance, India purchased two BWRs from the United States. They were built at Tarapur, north of Bombay, and use enriched uranium fuel. They provided the United States leverage over India's nuclear program but also became a source of considerable friction between the two countries after India's nuclear test in 1974.

[86] Strauss 1955.

proposals, the Duquesne Light Company was selected in November 1954. Duquesne agreed to invest $5 million in the development of the reactor, to provide the power plant site, and to build the turbine-generator.[87] It thereby became a partner of the nuclear power enterprise of Westinghouse and the AEC. It was the first U.S. utility to sign onto the U.S. civilian nuclear power program, and the result was the first commercial nuclear power plant in the United States, built in Shippingport, Pennsylvania. The overall project expenditure was to be about $85 million. The target date for completion was March 1, 1957.

The induction of a utility into the nuclear power program was both a financial and a public relations coup. Utilities were then among the most staid institutions in the country. They were monopolies regulated by state-run public utility commissions that guaranteed a rate of return on their investments. The assured rate of return meant that utilities were considered very safe investments on Wall Street. A utility signing on to nuclear power showed the world that nuclear power had come of age because utilities had the potential to both increase the access of the new technology to capital and to provide it with a market. The symbolism of a nuclear power plant being hooked into the grid of a utility also conveyed the strong message that the age of nuclear power was beginning.

A summary of the chronology of the early years of Rickover's reactor program, including the Shippingport reactor, is shown in Table 3 below.

Shippingport was not the first reactor to be on an electricity grid. The British beat the United States by bringing on line in October 1956 a dual-purpose reactor at Calder Hall, producing plutonium for the military and electricity for civilian use. And the Soviets had a small, 5 MWe graphite-moderated reactor at Obninsk connected to a grid in June 1954, over two years before the British. The first large Soviet power-producing reactor was a dual-use reactor that went on line in September 1958 at the Tomsk-7 nuclear weapons plant. The world's first nuclear electricity was generated in the experimental breeder reactor in Idaho in 1951.[88]

---

[87] Duncan 1990, p. 198; and Hewlett and Holl 1989, pp. 196-197.

[88] Makhijani et al., eds. 1995; Hewlett and Duncan 1990, p. 498; and for Soviet reactors, Thomas Cochran, Natural Resources Defense Council, personal telephone communication, April 5, 1996.

Table 3: Chronology Leading to the First U.S. Power-Generating Reactor[89]

1946:	Captain Rickover, head of Navy nuclear propulsion team, sent to Oak Ridge to learn about nuclear energy from scientists and engineers of the World War II Manhattan Project to build the atom bomb.
1947:	At Rickover's urging, Westinghouse establishes a separate atomic power division within the company (the Bettis Lab near Pittsburgh).
Spring 1947:	Plans set for General Electric's AEC-funded Knolls Atomic Power Lab near Schenectady, NY, for research. (General Electric does research on Redox, a reprocessing technology, to allow uranium recovery at Hanford.)
Fall 1948:	General Electric proposes an intermediate power breeder reactor.
December 10, 1948:	Westinghouse signs contract to build thermal submarine propulsion reactor (the Mark I), with work to be conducted at Westinghouse Bettis Lab, and the first land prototype to be built somewhere in the western United States.
February 18, 1949:	AEC approves selection of an Idaho site for a reactor testing station on Navy land.
1952:	Joint Committee on Atomic Energy and Senate pressures Navy to reverse a decision not to promote Rickover (which would have led to his mandatory retirement); Rickover becomes rear admiral.[90]
February 1952:	AEC authorizes a second land prototype submarine reactor, using General Electric's intermediate sodium-cooled reactor Mark A. This design is eventually installed (as Mark B) in a second nuclear sub, the *Seawolf*. (*Seawolf*'s sodium-cooled reactor was replaced by a PWR in 1958-59.)
March 1952:	Third land prototype naval reactor for aircraft carrier (CVR) is authorized by AEC; work by Westinghouse at Bettis Lab.
late 1952:	Westinghouse's Mark I land prototype ready for testing at Idaho.

---

[89]Unless otherwise noted, the references for this chronology are Hewlett and Holl 1989; and Duncan 1990.

[90] Pringle and Spiegelman 1981, p. 158.

1953	Joint Committee on Atomic Energy obtains money to continue funding CVR team (i.e., Navy-Westinghouse) as civilian power project (which became the Shippingport reactor). [91]
May 1953:	Pentagon cancels Westinghouse's aircraft carrier reactor (CVR project).[92]
July 9, 1953:	AEC assigns the civilian nuclear power project to Rickover.
December 1953:	AEC Commissioner Murray: "We recognize that the costs of power, derived from this first reactor, will be higher than costs from modern plants. But we will never really know the answer to costs until we build and operate several large-scale re-actors."[93]
September 1954:	Ground-breaking for Shippingport by Eisenhower.
January 1955:	Launching of *Nautilus*, the first submarine with nuclear propulsion. It was powered by a PWR.
December 1957:	Shippingport (60 MWe PWR) begins operation by Duquesne Light Company of the city of Pittsburgh (headquarters of Westinghouse).

Despite the Soviet and British head start with graphite-moderated, power-producing reactors, it was Westinghouse's PWR at Shippingport that shaped the future of nuclear power first in the United States and then in the rest of the world. The United States was far richer and more powerful than Britain. The Soviet Union had no global program comparable to "Atoms for Peace." Moreover, the graphite-moderated reactor design suffered a serious setback when one of Britain's military plutonium reactors, which was like the Calder Hall reactors in that it was also graphite-moderated, suffered a serious accident in October 1957,[94] about two months before Shippingport went on line. The earlier completion of the Soviet and British graphite-moderated reactors would not make a substantial contribution to nuclear power develop-

---

[91] *Nucleonics* 1953b

[92] *Nucleonics* 1953a

[93] *Nucleonics* 1953b

[94] Makhijani et al., eds. 1995, Chapter 8.

Shippingport Atomic Power Station (in service 1957-1982), Pennsylvania. The first commercial reactor in the U.S., Shippingport was built through a joint project between the Navy and Westinghouse, with the subsequent participation of the Duquesne Light Company. (Credit: Duquesne Light Company)

ment other than in the context of military plutonium production. Shippingport's greater economic significance for investors also lay in the fact that a private utility had contributed capital to and participated actively in project, whereas both the Soviet and British efforts were purely governmental.

For those directly involved in the construction and operation of the Shippingport reactor, however, it was far from clear that the technology would be economical. In November 1959, Charles H. Weaver, vice-president of Westinghouse's Bettis Lab, where Shippingport had been designed, argued for further development of the reactor core, noting that:

> A program limited to exploitation of existing basic technology will not meet the goal of reducing the cost of nuclear power to levels which will make it fully competitive throughout this

country and abroad.[95]

There were critics of nuclear power even for the naval vessels, where cost was a much smaller consideration, and where nuclear energy possessed strategic advantages over petroleum, given the premises on which the Cold War was fought. Still, during the early 1960s, there was debate in the Pentagon as to the desirability of a large number of nuclear ships in view of their cost. According to the history of the naval reactor program published by the Naval Institute Press, Rickover gave a briefing to naval officers and other officials in February 1962, explaining the special needs of nuclear reactor design that made it difficult to come up with a cheap reactor. He recalled the previous failures to build cheap diesel power plants for ships. In Duncan's paraphrase of Rickover's lecture:

> . . . the development of small, light, cheap components was often . . . unsuccessful. Nuclear propulsion was no different. Essential qualities such as ruggedness, reliability, and safety did not lead to small cheap reactors.[96]

Rickover wanted to make sure his reactors were a success in terms of performance. But ruggedness and reliability that were a requirement for naval reactors were also needed for civilian power plants, especially given the perilous consequences of accidents, some of which had already been investigated by the AEC by 1957 (see page 89). While Rickover's implicit argument that cost be relegated to secondary status was tenable for naval reactors, it did not apply to commercial power development, where competitiveness with coal-fired power stations was the decisive factor in whether nuclear power would be viable.

Other Skeptics

During the late 1940s and 1950s, the United States was undergoing a considerable transformation in its energy situation. Prior to and during World War II, the U.S. was virtually self-sufficient in petroleum. But the enormous growth in the number of automobiles in the decade, as well as the explosive growth of other uses of petroleum, resulted in

---

[95] Weaver 1959.
[96] Duncan 1990, p. 125.

the United States becoming a consistent net importer by the end of the 1940s. By 1960, the U.S. was importing almost one-fifth of its consumption. This trend was clearly evident in the 1950s. Moreover, it was occurring at a time when western Europe was also becoming highly dependent on imported oil. Imports of other resources were also growing, including strategic commodities like aluminum. One of the official reviews of the resource situation in the early 1950s was conducted by a commission appointed by President Truman, called The President's Materials Policy Commission. It came to be known as the Paley Commission, after its chairman.

In the energy sector, the prime area of concern that the Paley Commission addressed was petroleum. Its concluded in its 1952 report that there may be oil shortages by the 1970s. While it did not devote a great deal of attention to non-fossil fuel energy sources, its conclusions about them were as follows:

> Nuclear fuels, for various technical reasons, are unlikely ever to bear more than about one-fifth the load.

> We must look to solar energy. . . .

> Efforts made to date to harness solar energy economically are infinitesimal. It is time for aggressive research in the whole field of solar energy—an effort in which the U.S. could make an immense contribution to the welfare of the free world.[97]

The Paley Commission made an even stronger negative assessment of nuclear energy in its energy section of the report stating that "At this time, it does not appear that nuclear fission can be regarded as a contribution in any substantial degree to electric generation during at least the next 10 or 15 years, and the probability is that the atomic energy industry will remain a net heavy consumer of electricity," including presumably the consumption by the military sector.[98]

The Commission also encouraged work on wind energy and biomass. These three sources—direct use of solar energy, wind energy, and biomass—are exactly the energy sources that are likely to be the foundation of a renewable energy future. However, despite its evidently greater potential documented at the time by a presidential com-

---

[97] Paley Commission 1952, Vol. IV, p. 220.
[98] Paley Commission 1952, Vol. III, p. 39.

mission, a significant renewable energy effort was not made until the oil crisis was upon the U.S. in the 1970s; nuclear energy was vigorously pursued in the 1950s.

Given the assessment that nuclear energy could meet only a modest fraction of energy requirements at best, it seems illogical that nuclear energy was pursued vigorously rather than solar and other renewable energy sources. After all, the claims that nuclear fuel supply would stretch far into the future because of a potentially large physical resource base could also have been made for renewable energy sources. Evidently, they did not provide the same propaganda capital in the Cold War as did nuclear energy. Interestingly, a lack of government money was accompanied by a lack of corporate research effort and an absence of interest on the part of large numbers of scientists and engineers.

CHAPTER 5:
FROM "TOO CHEAP" TO BUST

The First Civilian Reactors

The Shippingport, Pennsylvania, plant did not unleash a flood of orders for nuclear power plants. In 1954, under pressure from the AEC and from the nascent nuclear industry, Congress had amended the Atomic Energy Act to encourage private industry to enter the nuclear energy field. However, the newly revised Atomic Energy Act, the desire of potential nuclear vendors like Westinghouse and General Electric to market light water reactors, and the technical success of Shippingport in generating electricity were not enough to convince utilities to take the nuclear plunge.

A range of subsidies and inducements were ultimately required before civilian nuclear power became a reality. These included relief from paying normal interest charges on publicly owned fuel and the use of AEC laboratories at no charge. One of the most significant subsidies, which continues to this day, was embodied in the Price-Anderson Act in 1957, which limited the liability of nuclear utilities and vendors in the event of an accident. Despite these subsidies, the AEC's director of reactor development, Kenneth Davis, found it necessary in 1957 to try "to bolster the sagging spirits of American industrial leaders, who were becoming increasingly disillusioned by the fading prospects for nuclear power."[99]

The economics of Shippingport did not help matters. The estimated cost of power from Shippingport was more than ten times the cost of conventional power—about 6.4 cents per kilowatt hour, compared to 0.6 cents for coal-fired plants.[100] A factor of ten higher cost might be

---

[99] Hewlett and Holl 1989, p. 410.

[100] Hewlett and Holl 1989, p. 421. Electric power costs are often measured in mills per kilowatt hour. One mill is equal to one-tenth of a U.S. cent. Thus, 6.4 cents per kilowatt hour of electricity is usually written as 64 mills per kWhe.

understandable for a pilot plant, but it did not warrant any commercial rush to the technology. Further, there did not seem to be early prospects for the dramatic declines in prices that were needed to make nuclear power competitive with coal.

It was government—mainly the AEC and the congressional Joint Committee on Atomic Energy—and the large nuclear contractors who were the main proponents of nuclear power at this stage, which remained more an ideological and a military than an economic issue. For G.E. and Westinghouse, commercial nuclear power offered the prospect of huge profits both from the nuclear navy and from civilian plants. The other large players, the utilities who would actually build, own, and operate the nuclear power plants, were more circumspect because they had much more at risk from an economic standpoint, and little to gain.

The diminishing fervor for nuclear power was made clear in a confidential 1957 memorandum to Strauss, which discussed the "moans and groans" of industry leaders about the expense of nuclear power development and the difficulty of the technical problems to be overcome. Some nuclear vendors were talking about dropping out of the nuclear business altogether, according to the report, and thought it "disgraceful" that the utilities were so reluctant to participate in arrangements to finance nuclear projects. The only way to save the situation, according to the report to Strauss, was for the chairman to use his prestige and lobby selected utility executives directly to persuade them that nuclear power was to their advantage.[101]

But lobbying by Strauss was not enough to overcome the economic and technical realities. Nuclear power was still an infant technology in regard to large-scale generation. A meeting of reactor experts in October 1957 concluded that "a long campaign of patient and painstaking development, rather than a dramatic technical breakthrough, was the likely road to nuclear power."[102]

Commerical nuclear power advanced slowly from the mid-1950s to the early 1960s, subsidized and pushed by the government, until 1962.[103] The following is a list of the start-up dates of the first several

---

[101] U.S. Atomic Energy Commission, Memo from Robert Zehring to Lewis Strauss, November 4, 1957, as cited in Hewlett and Holl 1989, pp. 416-417.

[102] Hewlett and Holl 1989, p. 414.

[103] Perry et al. 1977, p. 17.

U.S. civilian power reactors,[104] all of which benefited from some form
of government subsidy or supporting involvement:

1957	Shippingport, Pennsylvania (Westinghouse/AEC 200 MWt PWR; Duquesne Light)
1959	Dresden-1, Chicago, Illinois (General Electric 700 MWt BWR; Commonwealth Edison)
1962	Indian Point 1, Hudson River, New York state (615 MWt PWR; Consolidated Edison)
1963	Yankee-Rowe, Rowe, Massachusetts (600 MWt PWR; Yankee Atomic)
1964	Big Rock Point, Michigan (General Electric 240 MWt BWR; Consumer's Power)

In their quest to boost the nuclear power industry, General Electric
and Westinghouse guaranteed the utility companies very optimistic
generation costs, engaged in aggressive political lobbying, and funded
pro-nuclear public relations campaigns. In the words of historian and
economic analyst of the subject Steve Cohn, G.E. and Westinghouse:

> . . . priced thirteen nuclear plants at costs leaving them compet-
> itive with fossil-fuel generating facilities. Most conventional
> estimates project losses in the $800 million to billion dollar
> range, with per unit subsidies in the neighborhood of 50%.[105]

Cohn also notes that "Westinghouse (and to a lesser extent G.E.)
assumed multi-billion dollar, low-cost uranium supply contracts, which
subsequent uranium price increases forced Westinghouse to abro-
gate."[106]

Deliberate underpricing was meant to spur a huge rush of reactor
orders in a practically moribund market. Light water reactor technology was accumulating a list of safety and cost vulnerabilities. G.E. and
Westinghouse felt that they had to take action to establish the technol-

---

[104] Start-up dates are the ones on which the operating license for the reactor was
granted. The operating licenses were granted after the reactors were commissioned and
then run for some time so that they could be tested. Note that capacities are cited in
terms of thermal power, not electrical power. Electrical power was typically about 30
percent of thermal power. Source for start-up dates and capacities, NRC 1995,
Appendix A and Appendix B.

[105] Cohn 1990.

[106] Cohn 1990.

ogy in the marketplace before it became obsolete. As a General Electric vice-president later told *Fortune* magazine:

> We had a problem like a lump of butter sitting in the sun. If we couldn't get orders out of the utility industry, with every tick of the clock it became progressively more likely that some competing technology would be developed that would supersede the economic viability of our own. Our people understood this was a game of massive stakes and that if we didn't force the utility industry to put those stations on-line, we'd end up with nothing.[107]

The strategy worked. Utilities took up the manufacturers on contracts that offered fixed prices (other than adjustments for inflation) and power plants that would be delivered on a "turnkey" basis—all the utility would have to do would be to turn the key and the power plant would be hooked into its network, making electricity and money.[108]

Unrealistically optimistic cost projections by both AEC and industry, and "turnkey" solicitations deliberately underpriced to take some initial losses, led to what has been called the "Great Bandwagon Market" for nuclear plants in the U.S.[109] This phase of nuclear power development resulted in total sales of 45,000 MWe of nuclear power plant capacity by the end of 1967.[110] This means that about 45 percent of the capacity of all nuclear generation ever brought on line in the United States was ordered during this "Bandwagon Market" of the mid-1960s.

G.E. and Westinghouse lost money on these contracts, but they had

---

[107] Demaree 1970, p. 93.

[108] Industry memory of how LWR technology came to be established can be very defective. For instance, Alvin Weinberg, one of the foremost proponents of nuclear power, recalls that in 1985, "One publicist for the Atomic Industrial Forum claimed that the light-water reactor had been chosen after long and careful analysis because it possessed unique safety features. I knew this was untrue: pressurized water had been chosen to power submarines because such reactors are compact and simple. Their advent on land was entirely due to Rickover's dominance in reactor development in the 1950s; and once established, the light-water reactor could not be displaced by a competing reactor. This is not to say that such reactors are unsafe. . . . But to claim that light-water reactors were chosen *because* of their superior safety belied an ignorance of how the technology actually evolved." (Weinberg 1994, pp. 231-232.)

[109] Sporn cited in Bupp and Derian 1978, p. 49.

[110] Bupp and Derian 1978, p. 49.

firmly established light water reactors as a principal design, if not *the* principal design for commercial nuclear power in the United States. According to a RAND Corporation study by Perry et al., funded by the National Science Foundation:

> By mid-1966 it was obvious that for both turnkey and non-turnkey plants, bid prices considerably understated probable manufacturing costs. When the first 25 plants were finally completed, their capital investment costs were roughly twice what had been estimated when they were purchased.[111]

Perry et al. estimate that the actual cost of these plants was about $220 per kilowatt electrical, almost double that of the turnkey contract price which averaged about $113 per kilowatt electrical. Further, General Electric and Westinghouse losses were estimated to be between $875 million and $1 billion on the 13 turnkey contracts. Utilities that did not have turnkey contracts bore additional cost overruns.[112]

The first civilian reactor built during the "bandwagon" phase—without a direct government subsidy (other than the Price-Anderson Act), but with the builder taking a loss—was a 515 MWe General Electric Boiling Water Reactor (BWR) at Oyster Creek, New Jersey. The purchase was announced by Jersey Central Power & Light in December 1963, which justified this move stating that within five years of start-up, the plant would produce electricity more cheaply than any other generating system.[113] The principal trade journal of the nuclear industry hailed Jersey Central's analysis, saying that it "'confirmed in the strongest possible way' that earlier economic analyses were 'obsolete'," since the Oyster Creek project "established that the costs of nuclear power were 'now at levels which would have seemed incredibly low a year ago'."[114] Of course, the losses deliberately taken by the manufacturers on these early plants were not factored into such calculations.

---

[111] Perry et al. 1977, p. xii.
[112] Perry et al. 1977, pp. xii and 35.
[113] Bupp and Derian 1978, pp. 42-43.
[114] Bupp and Derian 1978, p. 43.

Safety

In the rush to build the new civilian reactors, safety was not as much of a consideration as it should have been. Cautious internal recommendations that power reactor should be developed slowly were cast aside. The most serious type of nuclear accident, the core meltdown, was considered in the 1950s. Yet experimental work on mechanisms by which loss-of-coolant accidents might occur and the full consequences of such accidents were elaborated on only in the 1960s and early 1970s, after LWRs were already established as the dominant commercial reactor type in the U.S. When safety considerations began to be factored in, costs escalated.

In 1947, the AEC appointed a Reactor Safeguards Committee to oversee safety issues regarding nuclear reactors. This committee was renamed the Advisory Committee on Reactor Safeguards in 1953. It was to play an important role in being an influential inside voice advocating greater caution and a slower pace of large-scale nuclear reactor development.[115]

The AEC also took other measures to assess the impact of reactor accidents. One of the most important studies done in the 1950s was one that became known by its report number: WASH-740. It was an assessment by Brookhaven National Laboratory of the radiological and other health, injury, and property damage consequences of a severe nuclear reactor accident. The reactor was assumed to be 500 MWt (100 to 200 MWe), located 30 miles upwind of a major city. WASH-740, published in March 1957, concluded that up to 3,400 people could die, up to 43,000 could be injured, and property damage could be as high as $7 billion (about $38 billion in 1995 dollars, using the consumer price index). Between 18 square miles and 150,000 square miles of area could be affected.[116]

A few months later, in September 1957, Congress passed the Price-Anderson Act limiting liability of nuclear power plant owners—that is, utilities—to $500 million dollars, plus an amount that could be required of the licensee that could be secured through private insurance. A total fund of $560 million per accident was to be established.[117] The

---

[115] Ford 1982, pp. 42 and 49.

[116] WASH-740 1957, pp. 13-14.

[117] Ford 1982, pp. 45-46.

Price-Anderson Act did not actually require nuclear power plant owners to acquire any private insurance from insurance companies as such.
Self-insurance or "other proof of financial responsibility" could be
deemed to be sufficient to meet the private insurance requirement. Not
only was the sum of $500 million less than 10 percent of just the
property damage costs estimated by the WASH-740 report, it would not
be fully devoted to compensation. The sum included "reasonable costs
of investigating and settling claims and defending suits for damage."[118] (The insurance was increased to $7 billion in 1988, when
Congress amended the Act.)[119]

Instead of letting Brookhaven's WASH-740 report and many
voices of caution slow it down to see how accidents might be avoided
and their costs reduced when they could not, the AEC plunged ahead
with the development of large-scale nuclear reactors in the 1950s and
early 1960s. In fact, the reactor-size postulated in WASH-740 was far
smaller than many of the reactors that were ordered in the 1960s. The
safety and cost consequences of that haste would come back to haunt
the nuclear establishment in the decades that followed.

1. Reactor Safety Basics

Like any other complex device, nuclear reactors have safety vulnerabilities and can suffer a variety of accidents. Since they contain a
large amount of highly radioactive material, some of the most intense
questions surrounding nuclear power have focused on the potential for
catastrophic accidents, the probabilities of such accidents, and the reactor design and engineered safety measures that can be taken to prevent them or to minimize their consequences, if they do occur.

There are a number of different kinds of events that could cause
considerable releases of radioactivity, although not all reactors are
equally vulnerable to all types of accidents:

- natural catastrophes such as earthquakes that could cause a failure of containment;

---

[118] Price-Anderson Act 1957.

[119] For a discussion of the Price-Anderson Act as a method of compensation, see
Berkovitz 1989.

- loss of control of a reactor in a manner that would cause a sudden, large increase in power output (called a reactor power excursion), as happened at Chernobyl, Unit 4 in 1986;

- loss-of-coolant accidents (see below);

- graphite fires in graphite-moderated reactors;

- nuclear explosions in some breeder reactor designs.

In addition to accident possibilities, there is also, the problem of potential terrorist sabotage. In this case also, not all reactor designs are equally vulnerable (see Part Two).

One basic issue in safety is whether the design of the reactor itself should be such as to obviate the possibility of catastrophic accidents. In Chapter 2, we noted that control of a reactor is fundamental to safety. Reactors depend on delayed neutrons to assure reactor control. If a reactor becomes prompt critical, control is lost and the sudden increase in power could destroy the reactor and its containment.

Similarly, a loss of coolant could result in the meltdown of fuel and potential release of radioactivity to the environment.

In theory, it appears possible to design reactors so that reactivity cannot increase beyond the delayed neutron fraction, so that prompt criticality is impossible. Similarly, in theory, it is possible to design reactors and containment structures so that catastrophic releases of radioactivity to the environment need not accompany a loss-of-coolant accident, but this requires that reactors be much smaller than typical power reactors. However, the approaches to design that yield assurance regarding protection from catastrophic releases of radioactivity also tend to make reactors very expensive, and therefore uncompetitive as a source of power. One cannot know with any practical assurance what the outcome might have been had a patient, safety-oriented reactor research and development strategy been adopted in the 1950s. As it happens, the rush to build commercial reactors resulted in the adoption of designs that have basic safety vulnerabilities.

All power reactor designs that have been deployed are to some extent vulnerable to loss of control. Reactors that are fueled in batches, such as LWRs, are more vulnerable to loss of control via control rod ejection especially at the initial stages of fuel loading than continuously fueled reactors, such as the heavy water CANDU reactor. The

NRC regards control rod ejection as having a very low probability. Of course, this does not mean that it cannot occur. For instance, loss-of-coolant accident probabilities were revised upward after the Three Mile Island accident.

Secondary containment can help greatly reduce radioactivity releases, provided the accident is not accompanied by an explosion that is violent enough to destroy it. The secondary containment of the Three Mile Island reactor prevented it from becoming the kind of lasting radiation disaster that was the result of the Chernobyl accident.

We will focus in this chapter on the loss-of-coolant accident because it is the most commonly considered catastrophic accident scenario for the most power common reactor type in the world, the LWR.

2. Light Water Reactors: Basics about Loss-of-Coolant Accidents

The basic problem with the conventional light water design is described by Lawrence Lidsky, an MIT nuclear engineer who is an advocate of an advanced reactor design. According to Lidsky,

> The design of a nuclear reactor requires three basic choices: the fuel, which undergoes the chain reaction; the moderator, which surrounds the fuel and facilitates the chain reaction; and the coolant, which carries the heat generated in the chain reaction off to do useful work. All the problems of the LWR stem from the particular choices that were made.[120]

In a conventional light water reactor, the fuel is in the form of a ceramic consisting of uranium dioxide. Ordinary water serves as both coolant and moderator. Heat generated in the reactor also converts water into the steam that drives the steam turbine. In PWRs, the steam is generated in a secondary circuit, while in BWRs, the water in the reactor itself is boiled to produce steam.

The ceramic fuel can withstand high temperatures, but it has very low heat conductivity and is very brittle. Because of its brittleness, the ceramic uranium fuel does not have sufficient structural integrity to maintain its form as the reactor heats and cools. Therefore, it is encased in hollow metal rods in the form of pellets. Because of the ceramic

---

[120] Lidsky 1987.

fuel's low heat conductivity, it must be made very hot to force the heat through the ceramic to the metal casing which is in contact with the water in the reactor that serves to keep the temperature below the fuel rod melting point.

This fuel design creates the difficulty that the operating temperature inside the fuel pellet is much higher than the melting point of the encasing metal rods. This means the fuel rods must be cooled to prevent them from melting. Thus, cooling water in an LWR not only serves the usual function in electricity generation of transferring energy from the fuel to the steam turbine, but it is also essential to maintaining the integrity of the reactor core itself. Even when the nuclear reaction has been shut down, there is often sufficient heat generated by the fission product decay (called "decay heat") in the fuel to raise the temperature above the melting point of the metal rod if continuous cooling is not provided (see Chapter 2). This does not present a problem as long as the metal is cooled by the surrounding water, which keeps the metal rod much cooler than the inner core of the fuel it contains. However, if cooling is lost, the situation can degenerate very quickly. In the absence of cooling, the decay heat alone is sufficient to damage a fuel rod in only five seconds, and completely destroy its integrity in 20 seconds.[121] Since all cooling is not likely to be lost instantly in a real-life situation, an accident that causes the water to boil or leak away is not likely to be that fast. But the fuel can melt in a matter of minutes after cooling water is lost or has boiled away.

In the worst case, the fuel core could fall to the bottom of the reactor vessel in a molten mass, melt through the reactor vessel, through the containment building surrounding the reactor, and into the ground below, where the intense radioactivity may be released to the general environment. This is the so-called "China Syndrome," in which the core is so hot that nothing could stop it, and it melts straight down, supposedly in the direction of China.[122] However, since the fuel would cool as it became mixed with external material, it would not actually go very far into the earth. In the worse case, it might reach the water

---

[121] Lidsky 1988, p. 220.

[122] This is a popular but geographically inaccurate phrase. Further, those who authored the phrase "China Syndrome" were considering only reactors in the United States. Accordingly, reactors in China might be presumed to suffer from the risk of the "U.S. Syndrome," which would also be geographically inaccurate.

table. To guard against this sort of accident, redundant, independent safety and backup systems are installed. Detailed quantitative studies of reactor designs based on assumed rates of component failure and other factors (called "probabilistic risk assessments") are used to attempt to quantify the risk and make it low, but there is always an element of uncertainty and such studies are difficult to verify. This approach to reactor safety is called "defense-in-depth," and it is intended to make the probability of a catastrophic loss of coolant accident very small.

A loss-of-coolant accident could occur if there were a break in a pipe carrying water into or out of the reactor vessel, leading to a sudden outflow of the water inside it. Malfunctioning of valves that control water flow could also cause a loss of coolant to occur. In such a case, the hot fuel would be uncovered, leading to various physical and chemical conditions that could cause a meltdown of the fuel. The Emergency Core Cooling System (E.C.C.S.) is designed to supply cooling water to the reactor core in case of the loss of its normal cooling water supply.

Reactor designers and builders try to make the probability of a loss of coolant small by a variety of means such as selection of materials, arrangement of piping and valving systems, and prescriptions for careful construction. But the probability of a loss-of-coolant accident cannot be reduced to zero. In other words, there is always the chance that a severe accident leading to a meltdown might occur. A partial meltdown is precisely what occurred in the Three Mile Island accident of 1979 in spite of redundant safety systems, some of which failed and some of which were simply over-ridden by confused operators who misunderstood what was actually happening. Luckily, in that instance, a total meltdown was avoided, and the fuel which did melt remained contained by the reactor containment vessel. Although the reactor core was essentially destroyed, the cooling water and containment building became heavily contaminated, and some radioactivity was released to the environment, an accident of catastrophic magnitude (such as the one which happened in 1986 in Unit 4 of the Chernobyl plant in the Soviet Union) was avoided.[123]

---

[123] For a discussion of the Three Mile Island accident, see TMI Commission 1979; and Jaffe 1981. It should be noted that the Chernobyl reactor is not an LWR. The Chernobyl accident was not initiated by a loss of coolant, but by explosions and fires resulting from an uncontrolled power surge.

Metropolitan Edison Company's Three Mile Island Nuclear Station, Unit 1, located along the Susquehanna River about 10 miles south of Harrisburg, Pennsylvania, as seen under construction. Unit 2 of this plant had a partial meltdown accident in 1979. (National Archives)

A complicating vulnerability of the LWR is a chemical characteristic of the metal fuel rods, which are made of an alloy (called zircaloy) of which the principal constituent is the element zirconium. Above a certain temperature, zirconium reacts with steam, resulting in the generation of hydrogen gas (along with zirconium oxide) and the generation of more heat. Under loss-of-coolant accident conditions, the fuel is uncovered but the rods are in contact with steam from water that has boiled. This can result in the accumulation of significant quantities of hydrogen that poses two potentially serious problems. First, since hydrogen is a non-condensable gas, it takes up space in the primary coolant system and can make it difficult to re-establish full coolant circulation even if the initial disabling problem is overcome and cooling systems are restored. And second, it creates the potential for a dangerous hydrogen explosion.

This chemical process occurred during the Three Mile Island accident. According to the President's Commission on Three Mile Island:

In the first 10 hours of the TMI accident . . . enough hydrogen
was produced in the core by a reaction between steam and the
zirconium cladding and then released to containment to produce
a burn or an explosion that caused pressure to increase by 28
pounds per square inch in the containment building.[124]

Fortunately, a large hydrogen explosion did not occur and the secondary containment was not breached. Millions of curies of radioactive
iodine-131 were retained within the containment building (in containment water and air), while about 13 to 17 curies of iodine-131 were
released to the atmosphere. In addition, 2.4 million to 13 million curies
of noble gases were released.[125]

3. Historical Aspects of the Light Water Reactor Accident Debate

The consequences of a loss-of-coolant accident were first officially
reported in the 1964-65 update of AEC's WASH-740 report. Daniel
Ford of the Union of Concerned Scientists, who chronicled this safety
issue in detail in his book *Cult of the Atom*, noted that the AEC was
"extremely reluctant to provide the detailed results of that study to the
A.C.R.S. [Advisory Committee on Reactor Safeguards],"[126] which was
charged by law to review the safety applications of every new nuclear
power plant application.

By the mid-1960s, the ACRS had become concerned that the size
of power plants was increasing too fast and that the outstanding safety
concerns were not being resolved prior to the start of construction. In
one case, the proposed site of a large nuclear reactor, Indian Point 2,
four times the size of any plant built up to that time, was only 35 miles
from New York City; 10 percent of the population of the United States
lived within a 50-mile radius of the site. According to Ford, "A.E.C.
regulatory staff engineers who reviewed the safety devices at Indian
Point used to joke half-seriously among themselves that Buchanan,
New York—which was north of Croton-on-Hudson and Hastings-on-
Hudson—should be renamed 'Hiroshima-on-Hudson,'" not because of
the danger of a nuclear explosion, but because of the danger of a cat-

---

[124] TMI Commission 1979, p. 30.
[125] TMI Commission 1979, p. 31.
[126] Ford 1982, p. 89.

astrophic spread of radioactivity from a reactor core meltdown.[127]

The ACRS knew about the WASH-740 update; indeed, one of its members, David Okrent, had a part in directing it. The AEC considered various ways of covering it up, but then decided under pressure to share it with the ACRS, which requested a briefing. The AEC provided the briefing during the May 5-7, 1966, meeting of the ACRS. Calculations indicated that a core melt in a large reactor might not be contained, and could be hot enough to melt through the concrete floor of the containment and into the earth. According to a former member of the ACRS, "This appears to be the first unequivocal statement by the regulatory staff to the effect that containment failure was a likely result of core melt in large LWRs."[128]

The analysis of the potential cause of loss-of-coolant accidents and their possible consequences did not reassure the ACRS. Early reactors had no emergency system to cool the reactor core in case of a loss of coolant. Newer reactors, which were far larger, were required to have emergency core cooling systems, but these were "something of an afterthought—additional pumps and piping that was tacked onto the plants at the last minute. The A.C.R.S. was unsure that the untested E.C.C.S. [Emergency Core Cooling System] . . . would work satisfactorily and felt that much more research on the subject was needed." It also wanted an additional back-up system to prevent a molten core from melting right through the reactor building, in case the emergency core cooling systems did not work. Industry did not think the additional safety measures were needed.[129]

The congressional Joint Committee on Atomic Energy was a participant with the AEC in pushing the reactor program ahead with little effort to ensure safety. This attitude persisted for decades. For instance, in 1968, Congressman Craig Hosmer, a senior Republican member of the Joint Committee on Atomic Energy who was unhappy with the ACRS, received a standing ovation at a nuclear industry meeting when he suggested "burning the Advisory Committee on Reactor Safeguards at the stake."[130]

During the mid-to-late 1960s there were numerous studies, exper-

---

[127] Ford 1982, p. 89.

[128] Okrent 1981, p. 102.

[129] Ford 1982, p. 91.

[130] Congressman Hosmer, quoted in Okrent, p. 185.

iments, and reviews of the problem of a loss of coolant and the ability of the emergency core cooling system to replace lost cooling water in the event of an accident. Analyses and experiments at Oak Ridge National Laboratory indicated that, due to swelling of the fuel rods, steam pockets, and other problems, the emergency core cooling system might not be able to do its job of keeping the fuel rods under water. One response of the AEC was to try to suppress discussion of the work; another was to cut off funding for the research.[131]

More than one problem discovered during this period came back to haunt the industry in the Three Mile Island accident, where there was a loss of coolant and a partial meltdown. There was a runaway reaction between the zircaloy cladding of the uranium fuel rods and the hot steam, which made the rods brittle and released hydrogen.

These concerns became public knowledge during a marathon hearing process that began in January 1972. As the safety problems became better known and as more experiments were carried out at the AEC's Oak Ridge and Idaho laboratories, public concern grew and there were more court challenges to the AEC's licensing of new reactors. To fend off these challenges, the AEC decided to hold a single rule-making hearing on the Emergency Core Cooling System and its ability to prevent a meltdown in the event of a loss-of-coolant accident.[132]

The AEC's position was that there was no problem. But there were now some scientists outside the government who had analyzed the core cooling problems. Henry Kendall and James MacKenzie of the Union of Concerned Scientists were foremost among them. They challenged the official public positions with their own analyses. Many scientists and engineers within the nuclear establishment, notably at Oak Ridge National Laboratory and at Idaho National Engineering Laboratory, agreed with the negative conclusions of the independent scientists.[133]

By the end of these hearings the AEC's credibility was seriously compromised. They showed that there was an inherent conflict of interest in the AEC, which was a promoter of nuclear power and the owner and builder of the country's nuclear weapons plants on the one hand and the watchdog of their safety on the other. The stark exposure of this institutional problem during the hearings on loss-of-coolant ac-

---

[131] Ford 1982, p. 98.
[132] Ford 1982, p. 116.
[133] Ford 1982, pp. 111-113 and 127.

cidents and the changes in the global energy scene in 1973 resulted in the 1974 congressionally mandated split-up of AEC. Congress created the Nuclear Regulatory Commission (NRC) and the Energy Research and Development Agency (ERDA). In 1977, Congress transformed ERDA into a cabinet level agency, the Department of Energy.

The split in the AEC gave to the NRC the role of regulating the nuclear industry. ERDA (later the DOE) would do research, development, and promotional activities for the civilian energy sector and at the same time carry on the nuclear weapons production functions of the AEC.

Public confidence in the AEC had been eroded by secrecy, incompetence, suppression of critical information, and the rise of the environmental concerns. It was a period when confidence in government was declining rapidly in other arenas, due in part to Watergate and the Vietnam war. Three Mile Island confirmed some of the public's fears, showing that very serious accidents were possible and that the U.S. had been close to a radioactive catastrophe.

4. Sodium-Cooled Fast Breeders

The other principal reactor design proposed for commercial nuclear power in the United States was also the subject of a behind-the-scenes tussle over safety. This was the sodium-cooled fast breeder reactor, which posed a number of hazards that were well understood even in the 1950s. The main concerns were the potential for a nuclear explosion due to the large quantity of plutonium in the core, a sodium fire (since liquid sodium burns on contact with air) or a sodium explosion (since liquid sodium explodes on contact with water). Fires or explosions could result in the spread of catastrophic quantities of plutonium and fission products.

Despite these potential problems, the breeder had been an early favorite of those who saw in nuclear power the promise of an endless energy supply. This was because it held the promise of greatly extending uranium resources by converting non-fissile uranium-238 into fissile plutonium-239. The world's first nuclear electricity was generated in 1951 in an experimental breeder reactor in Idaho (called EBR I).[134] General Electric had done research at its Knolls laboratory with the

---

[134] Lamarsh 1983, p. 138.

idea of building commercial breeder reactors, but concluded after considerable expenditures that the intermediate energy neutrons that it had chosen for the reactor were not suitable for breeding.[135] However, the U.S. Navy purchased, built, and installed a G.E. sodium-cooled reactor in the naval submarine *Seawolf* in 1956. (The reactor was replaced in 1958-59 by a PWR.)

While the *Seawolf* reactor was being constructed, a consortium of utilities led by Detroit Edison proposed to build a sodium-cooled fast breeder reactor, to be named after the lead scientist in building the first nuclear reactor in the Manhattan Project, Enrico Fermi. The proposed site was Lagoona Beach on Lake Erie between Detroit and Toledo. Detroit Edison was at the time headed by Walter Cisler, an influential proponent of the private development of nuclear power, who nonetheless wanted the government to insure privately owned nuclear reactors in case of catastrophic accidents.[136]

The AEC gave its approval to the project in August 1955. But in June 1956, the ACRS recommended against proceeding with the sodium-cooled reactor because the theoretical calculations and very limited experiments on which predictions of safety were based had not been confirmed. The ACRS was particularly concerned about unresolved safety issues because the plant was to be built so close to a heavily populated city.[137]

Some of the reservations about the safety of sodium-cooled fast reactors stemmed from the fact that there had already been a partial meltdown in the EBR-I breeder reactor in Idaho in November 1955. This occurred during a test of whether the reactor could withstand a power surge. The experiment revealed "that under certain conditions

---

[135] Pigford 1996.

[136] In 1954, during the congressional debates regarding the revision of the 1946 Atomic Energy Act, Cisler had lectured Congress to give private industry the opportunity to "to perform its natural function of seeking out economic methods of utilizing this natural energy resource and making the resulting benefits available to all in a normal manner" and warned it against restricting "opportunities by continuation of existing law." But in 1957, during consideration of the Price-Anderson Act, Cisler was in favor of government protection from liability: "The absolute necessity of insurance against a catastrophe involving the extensive public liability, in adequate amount, cannot be overstressed." Walter Cisler, as quoted in Mazuzan and Walker 1984, pp. 29 and 103.

[137] Ford 1982, pp. 54-57; and Mazuzan and Walker 1984, Chapter V.

an increase in core temperature produced a rise in reactivity, which could lead to a power runaway and core meltdown."[138] A two-second misunderstanding regarding the control of the reactor during the experiment resulted in the partial meltdown.[139]

The AEC overruled the ACRS report and its negative recommendation, while paying lip-service to the concerns raised in it. The United Auto Workers and two other unions filed suit to block the reactor. They prevailed in the appeals court, but were defeated in the Supreme Court, which ruled that Congress had allowed the AEC to make licensing and siting decisions.[140] Secrecy, the ability to override democratic procedures, and a very broad decision-making charter given to the AEC, primarily intended for the nuclear weapons program, were now used with a heavy hand in the civilian nuclear power program.

The Fermi I reactor, which was basically a scale-up of EBR I, was completed in 1963. It was still in the testing phase required to obtain an operating license when it suffered a partial meltdown accident in 1966 due a partial cooling system blockage by a piece of zirconium metal. It had not reached full power even once since its 1963 start-up. It was shut down until 1970, when it was reopened for further testing. It was permanently shut in 1972.[141] Fortunately, the quantity of fission products in the reactor at the time of the accident was not large, since the reactor had generated very little power. The radioactive materials that escaped the reactor core remained within the secondary containment.

Cost

1. 1950s to the 1970s

Figure 6 on the next page shows the evolution of estimated costs of nuclear power in the first two decades. Note that this graph represents the *advertised* cost of nuclear power from plants ordered in that year.

---

[138] Hewlett and Holl 1989, p. 352.
[139] Mazuzan and Walker 1984, pp. 127-128.
[140] Mazuzan and Walker 1984, Chapter VI.
[141] Alexanderson, ed. 1979.

Figure 6: Nuclear Power Costs

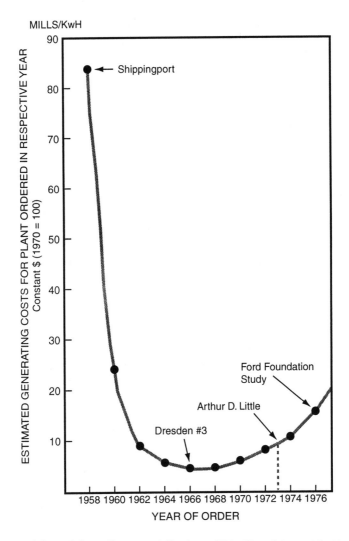

Source: Adapted from Bupp and Derian 1978, Fig. 5.2, p. 95. Used by permission of Basic Books/HarperCollins.

Some of the lowest costs on this graph turned out to be serious underestimates. The first major cost increases were due to safety considerations. As Bupp and Derian noted in 1978, a year before the Three Mile Island accident:

> The blunt fact is that 15 years after the Oyster Creek sale, light water technology has still not attained the technical maturity which its promoters thought it had reached years earlier. . . . There is a continuing disagreement over "how safe is safe enough?" In fact, by the mid-1970s it was clearly this dispute which was at the center of the economic woes of nuclear power.[142]

Between 1955 and 1965, according to Bupp and Derian, the increase in size of power plants gave rise to "classic 'economies of scale,'" drastically reducing materials requirements per unit of power. In 1955, a 180 MWe reactor had needed "more than 30 tons of structural steel and about one-third of a cubic yard of concrete per megawatt." In 1965, the materials requirements per unit of power for a 550 MWe plant were less than half those amounts. But by the mid-1970s, these classic economies had disappeared. Reactor size had increased still more, to 1,200 MWe, but the materials requirements per unit of power had gone back up to the levels used in a 200 MWe to 300 MWe 1960 reactor. Higher materials requirements were the result of more stringent safety requirements for power plant licensing compared to the earlier period.[143]

Increasing materials requirements, more stringent materials specifications, and other safety requirements, including redesign of some aspects of the power plants, led to an increase in safety, but also an increase in costs. Figure 6 shows how, after an initial sharp fall, costs began to rise again toward the end of the 1960s, in part due to safety requirements. A decade later the Three Mile Island accident revealed even more safety problems to be fixed. When these were mandated by the NRC, nuclear power costs increased even more (see below).

Figure 6 shows that the only period in which nuclear power plant costs were relatively low for utilities was from about the early 1960s to the mid-1970s. During this period, the cost of reactors was kept artificially low by three factors:

- inadequate consideration of safety requirements;

- government subsidies (until 1962);

---

[142] Bupp and Derian 1978, p. 156.
[143] Bupp and Derian 1978, pp. 156-157.

• industry decisions to sell reactors below cost.

Costs escalated rapidly during the 1970s as the effect of these factors declined. Komanoff has calculated that between 1971 and 1978 (the year before Three Mile Island) nuclear power plant capital costs of completed nuclear power plants increased at a compound annual rate of 13.5 percent in real terms, for a total of 142 percent. Capital costs of coal-fired power plants, by contrast, increased 68 percent, including the costs of sulfur dioxide scrubbers.[144]

After Three Mile Island, expensive modifications to existing reactors were required. Opposition grew, as did delays. Further, rising electricity and energy prices had dramatically slowed the growth rate of the electric power industry. Interest rates had risen dramatically in the global financial crisis of the late 1970s and early 1980s, which greatly increased the economic penalty of delays. The effect was disproportionately large on nuclear power plants relative to fossil fuel plants because of the relatively higher capital costs of nuclear reactors.

2.   *1980s and Early 1990s*

The cancellations of reactors already on order or under construction began in the late 1970s and turned into a definitive trend in the 1980s. No new reactors were ordered after 1978 and many reactors that had been ordered before that date were canceled. There were two basic reasons for the cancellations. First, the growth rate of electricity relative to Gross Domestic Product (GDP) went down during the 1970s. The ratio of the growth rate of electricity to that of GDP had been two to one from the 1950s until about 1973. In the mid-1970s, it declined abruptly to about one to one. Utilities were ill-prepared for this sudden shift and failed to reduce the number of power plants under construction in a timely way. In many cases, hundreds of millions of dollars or even billions of dollars of costs had already been incurred by the time unneeded plants were canceled.

The second factor in rising nuclear plant costs was interest rates. In 1979, President Carter appointed Paul Volcker to head up the Federal Reserve with the immediate task of shoring up the value of the dollar, which was collapsing on international markets during the sec-

---

[144] Komanoff 1981, p. 2.

ond oil crisis of the decade (this time occasioned by 1979 Iranian revolution). Volcker did so by drastically raising interest rates. Since nuclear power plant costs are dominated by capital costs, rising interest rates affected them disproportionately. While fossil fuel prices increased in 1979, they began to decline gradually in the early 1980s, and then fell abruptly with the fall of oil prices in 1986. In contrast, interest rates in real terms stayed high during most of the 1980s. The predicted shortages of oil and natural gas did not materialize; large new resources of both were discovered around the world. Finally, the increases in oil prices in 1973-74 and in 1979 also caused a switch from oil to lower-priced coal resources. U.S. electricity generated from oil declined from 289 billion kilowatt hours in 1975 to 117 billion kilowatt hours in 1990, while coal-fired generation increased from 704 billion kilowatt hours in 1970 to 1,557 billion kilowatt hours in 1990. This shift enabled fossil fuel generation to cushion the impact of oil price increases. Finally, the availability of natural gas in ample quantities and at moderate prices coupled with the relatively low capital cost of natural gas-fired plants also negatively affected the prospects of nuclear power.[145]

Given the relatively high costs of nuclear power, public utility commissions under pressure from consumer and environmental groups began to disallow routine factoring in of all capital costs of new plants into the utilities' rate base. This factor and losses due to canceled nuclear plants made the financial condition of many utilities less stable and in some cases precarious. The largest utility bond default in history involved a failed nuclear power project in the west, called the Washington Public Power Supply System (WPPSS, popularly pronounced Whoops). Nuclear power lost favor with investors.

Problems also began to emerge in the 1980s that led to increasing operating costs for existing reactors. Increases in maintenance costs and problems with key pieces of equipment were among the major factors. One of the most troublesome problems at some plants was leaks in steam generators of pressurized water reactors, necessitating repair or replacement. Steam generators are the heat exchangers in which the pressurized water coolant from the reactor gives up heat to water in the secondary circuit, converting the latter into steam to drive the turbine of the power plant (see Figure 1 in Chapter 2).

---

[145] At least one commentator has concluded that low prices of natural gas are "the major deterrent to commercial nuclear power." (See Pigford 1996.)

Costs of nuclear power became unpredictable even for existing power plants due to the erratic nature of repair and maintenance problems. For plants completed in the 1960s and 1970s, power production costs were low if they did not experience serious problems with equipment repair or replacement. For other utilities, costs escalated. By the early 1990s, even operating costs had escalated greatly for many nuclear utilities. Even this aspect of high operating and maintenance costs had been forecast in December 1950 by Philip Sporn, president of the American Gas and Electric Service Corporation and chairman of the Advisory Committee on Cooperation Between the Electric Power Industry and the Atomic Energy Commission. He noted that, while low fuel cost would be an advantage of nuclear power, capital costs would likely be high and that "there does not seem to be any doubt that operation and maintenance of a nuclear reactor will be more expensive than the corresponding part in a conventional plant . . . "[146]

Rising decommissioning cost estimates added to investor and rate-payer concerns. Premature reactor retirements have begun to occur in the 1990s.

Utilities had traditionally been a safe investment. But now they had become risky for a host of factors, including changing technology, high operating and decommissioning costs, and deregulation. One large Wall Street investment firm, Shearson Lehman, estimated in a 1993 report on a conference held by Lehman Brothers that "as many as 25 of them [commercial nuclear power plants in the U.S.] could face premature shutdown in the next several to 10 years" due to high costs and competition.[147] During the period since 1986, operating and maintenance costs of nuclear power plants have been about three to four times those for coal-fired power plants, excluding fuel costs. Even when the far higher fuel costs for coal are added, the costs of producing nuclear power have been higher than coal-fired electricity every year since 1987 (inclusive), according to data compiled by the NRC.[148]

Wall Street, Lewis Strauss's point of departure for Washington, had joined the ranks of the nuclear skeptics.

The lack of progress on a nuclear waste repository added another

---

[146] Sporn 1951. His December 1950 speech was printed in February 1951 in *Nucleonics*.

[147] Shearson Lehman 1993.

[148] NRC 1995, p. 22.

negative element to the problems of nuclear utilities and vendors. In 1982, the U.S. government created a nuclear waste fund under the Nuclear Waste Policy Act. This law required nuclear utilities to pay one mill per kilowatt hour of nuclear-generated electricity to defray the costs of the repository program. It also set a date for the opening of a repository of 1998, which was by far the most ambitious schedule of any country in the world.

By the mid-1990s, the DOE had spent $4 billion of the fund and had little to show for it but delays, questionable science, and an escalating lack of public confidence. Space for storage of spent fuel was running out at more and more reactor sites. These utilities faced added costs of building on-site storage in addition to paying into the nuclear waste fund. They reacted to the mounting wastes at reactor sites by pushing for centralized interim storage, called Monitored Retrievable Storage (MRS), to be built either by the federal government in Nevada or by the utilities themselves on Native American land belonging to the Mescalero tribe in New Mexico, among other proposals. Nuclear power vendors reacted to the crisis by trying to sell more reactors abroad and by looking again to the military connection to provide them with cash. The MRS proposal on the Mescalero reservation was eventually rejected by the tribe. A similar proposal for a private MRS on land belonging to the Goshute tribe (Skull Valley) is still active, but faces resistance.

CHAPTER 6:
RADIOACTIVE WASTE[149]

Power plant radioactive waste management has crucial security and environmental implications. This is because the fission process generates materials that can be used, after reprocessing, in nuclear and radiation weapons and because some components of these wastes will endure for hundreds of thousands of years.

A huge quantity of very dangerous radioactive waste and large amounts of materials usable in nuclear weapons have already been generated by the world's nuclear power plants. They supply only about 15 percent of the world's electricity. The world's power plant spent fuel and reprocessing waste combined contain on the order of 100 billion curies of radioactivity.

Commercial spent fuel also contains about 800 metric tons of plutonium, enough, if separated from fission products and uranium, to make on the order of 100,000 nuclear warheads. Another 195 metric tons have already been separated from it, enough for 24,000 warheads.[150] The need to manage these wastes so that they remain secure and isolated from the human environment for periods of time that are far longer than the duration of any human institution, and indeed far longer than civilization itself, is a daunting and unprecedented task. The United States has about 30 percent of the world's spent fuel.

---

[149] This chapter is based on Makhijani and Saleska 1992, unless otherwise noted.

[150] Plutonium produced specifically for nuclear weapons uses contains 93 percent or more plutonium-239. Plutonium from commercial power reactors contains far larger quantities of higher isotopes and hence more of it is therefore needed to make a nuclear weapon. See Appendix C.

Radioactive Waste Basics[151]

Light water reactors are fueled with low-enriched uranium (see Chapter 2). The amount of radioactivity in nuclear waste created by commercial nuclear power plants is very nearly proportional to the quantity of electricity generated because well over 99 percent of the radioactivity is in the fission products created in the process of nuclear fission. The half-lives of these fission products vary from a small fraction of a second to millions of years, so that the radioactivity in the fuel rods declines over time, very rapidly at first and then more slowly. Spent fuel rods once withdrawn from a reactor are also thermally hot due to the heat produced by radioactive decay of the fission products. Figure 7 shows the radioactivity and thermal heat generation over time of one metric ton of spent fuel discharged from a PWR.[152]

A 1,000 MWe PWR operating at two-thirds capacity for one year discharges about 25 metric tons of spent fuel. The amount of radioactivity in this spent fuel, measured one year after discharge, is on the order of 70 million curies. This declines to about 10 million curies at the end of 10 years of storage. By comparison, the annual radioactive waste generated in all other reactor-related processes is about 2,000 curies for PWRs and about 11,000 curies for BWRs. About two-thirds of the radioactivity in other radioactive waste discharged by PWRs and about 80 percent of it for BWRs can be attributed to the reactor parts that become radioactive during reactor operation.[153] Most of the rest is in resins used to filter primary coolant water and the water in the pools where spent fuel is stored.

---

[151] The numbers in this section relate to waste generated by light water reactors. The picture for other reactor types is broadly similar, but the exact figures vary by reactor type and operating history.

[152] The weight of fuel in reactor is generally measured in terms of the amount of heavy metal in the fuel—that is, the weight of all other materials, such as the cladding material or the oxygen in the uranium dioxide fuel, is excluded. Fuel weight may be cited as metric tons of initial heavy metal (MTIHM) loaded into the reactor, as shown in the figure, or simply as metric tons of heavy metal (MTHM). For practical purposes, commercial reactor heavy metal is the sum of uranium and plutonium isotopes in the fuel, since the weight of other heavy metal elements is small.

[153] As noted in Chapter 2, one of the disadvantages of the BWR design is that radioactive primary coolant water comes into contact with more parts of the power plant that in the PWR. Therefore, the amount of radioactive waste from decommissioning BWR reactors is far larger than for PWRs of the same size.

**Figure 7: Radioactivity and Thermal Heat Generation—
One Metric Ton of PWR Fuel (Heavy Metal Basis)**

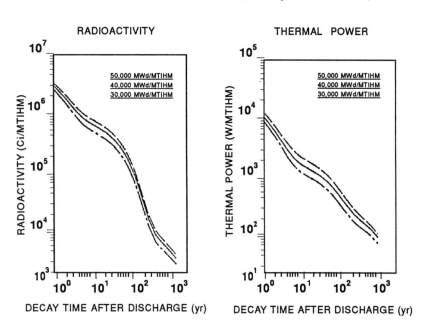

Source: DOE 1995, p. 29.

The story of the volume of wastes is quite different. By far the largest volume of waste created is in the uranium mine wastes and uranium mill tailings, which are disproportionately on Native American lands.[154] Depleted uranium, uranium-processing waste, and "low-level" wastes from routine operations constitute most of the volume. Table 4 on opposite page shows the volume and radioactivity content of wastes generated in the U.S.

---

[154] Tribal and colonial peoples have borne a disproportionate amount of the negative environmental impacts from uranium mining and milling throughout the world. See Makhijani et al., eds. 1995, Chapter 5.

Table 4: Volume and Radioactivity of Wastes in the Untied States by End of 1994

Waste Category	Volume (Cubic Meters)	Radioactivity (Million Curies)	Comments
Commercial spent nuclear fuel	12,000	26,600	
Reprocessing high-level waste	378,400	959	Military
Transuranic waste	more than 247,000	more than 2.67	Buried TRU and contaminated soil are not well characterized (military)
Uranium mill tailings	119,000,000	on the order of a few hundred thousand curies of thorium-230 and radium-226	IEER estimate
Low-level waste	4,600,000	23.5	
Mixed wastes	166,000	not available	Commercial mixed waste information not available
Commercial reactor decommissioning waste	27,000	not available	

Source: DOE 1995, Table 0.3, except for radioactivity in uranium mill tailings, which is an IEER estimate. (DOE does not compile data on uranium mine wastes.) They are of the same order of magnitude in volume as uranium mill tailings, but have lower radioactivity content. Numbers are rounded.

1. Waste Categories

The management requirements for short-lived radionuclides with half-lives on the order of a year or less, which are intensely radioactive for a brief period, are considerably different from those materials

Disposal of low-level radioactive waste at the Hanford Works in Washington State in mid-1970s. Radioactive waste management in the U.S. has been complicated by a poor classification system, based not so much on longevity and hazard, as on the process which created the waste as well as by poor disposal practices. (Credit: Battelle Northwest Laboratory)

which have long half-lives (10 to 15 years or more).[155] The short-lived radionuclides decay away in a decade or less, and monitored storage is sufficient as a management technique.

Long-lived radioactive materials present a different kind of problem, because their indefinite storage above ground for long periods is impractical from the point of view of guaranteeing their isolation from society and the environment. Wastes like spent fuel that contain weapons-usable materials pose especially grave problems of long-term management. At the same time, estimating the effects of disposal hundreds, thousands, or hundreds of thousands of years from the present is a matter that is replete with uncertainty. As a result, the long-term management of long-lived radioactive materials has become a daunting technical, economic, social, political, and military problem.

The inherent technical difficulties of radioactive waste management have been greatly exacerbated in the U.S. by poor regulations for waste classification and disposal, which are not so much based on longevity and hazard of the waste, as on the process that gave rise to the waste. Most countries have three classes of waste: low-level, intermediate-level, and high-level, defined according to the amount of radioactivity per unit weight or volume. The scientifically deficient nature of waste classification in the U.S. can be discerned from Table 5 below, which shows the average radioactivity per unit volume in various kinds of wastes in the United States.

The problems of management arising from this regulatory hodge podge have been compounded by past poor disposal practices, such as shallow land burial of radioactive waste in cardboard boxes. We have discussed this issue and the need for a sound waste classification system as the foundation of public policy regarding nuclear waste management in another book and hence we will not dwell on it here. The central recommendation of that book regarding waste classification has acquired considerable public support since its publication, but as yet no official action. We refer the reader to our earlier work for further details.[156]

---

[155] The specific activity (that is, radioactivity per unit weight) of a radionuclide is inversely proportional to its mass number (the number of protons plus neutrons in the nucleus) and its half-life.

[156] Makhijani and Saleska 1992, Chapters 4 and 5.

Table 5: Some Radioactive Waste Characteristics

	Average Concentration (Curies/ft 3)	Selected Samples (Curies/ft 3)	
Low-level Waste:			
Class A	0.1		
Class B	2	4.4	(NY Cinctichem)
Class C	7	160	(NY reactor average)
Greater than C	300 to 2,500*		
Transurantic Waste			
Contact-handled	0.57		
Remote-handled	47		
Military HLW	100	920	(Savannah River
		3.7	sludge)
			(Hanford salt cake)
		7,110	(SRP Glass, projected)
Commercial Spent Fuel	73,650**		

Source: Makhijani and Saleska 1992, Table 4, p. 26.
* The 300 figure is based on the 1985 inventory. The higher figure represents anticipated inventory in 2020, including some decommissioning
wastes.
** Based on average activity in all spent fuel at the end of 1989 and on
overall fuel assembly dimensions.

2.   Quantities of Weapons-Usable Materials

Both uranium-235 and the plutonium can be used to make nuclear
weapons. But the uranium used as reactor fuel in commercial reactors
is either natural uranium or low-enriched uranium, with the concentration of uranium-235 generally being below 5 percent. In practice, uranium enriched to at least 20 percent is required to make nuclear
weapons because an explosion requires that the mass of fissile material
not only become critical, but that it quickly become prompt critical. As
the fuel is used up, the amount of uranium-235 in a reactor declines.
Spent fuel from a light water reactor typically contains on the order of
1 percent uranium-235. This uranium cannot be used in nuclear weapons without being enriched to greatly increase the proportion of ura-

nium-235. Power reactors also create plutonium. A fraction of the uranium-238 (typically under 2 percent) is transformed into plutonium isotopes during reactor operation. Roughly half of this plutonium is fissioned in the reactor before the fuel is withdrawn; the other half remains in the spent fuel. Light water reactor spent fuel typically contains about 0.9 percent plutonium isotopes. Plutonium-239 constitutes about 60 percent of the plutonium isotope mix in a light water reactor (see Appendix C). It should be noted that in addition to plutonium-239, another isotope in the mix—plutonium-241—is also fissile. For reporting purposes, the quantities of fissile isotopes of plutonium are added together under the designation of plutonium-fissile (or Pu-fissile). Typically, LWR spent fuel contains about 0.7 percent fissile plutonium. Contrary to the propaganda of some of the enthusiasts for the use of plutonium as a reactor fuel, plutonium separated from power reactor spent fuel can be used to make nuclear weapons. This was demonstrated in a 1962 nuclear test at the Nevada Test Site. It takes about 40 percent more reactor-grade plutonium to make a nuclear warhead than plutonium made explicitly for such use, which is called "weapon-grade" plutonium. This is because plutonium from power reactors, called reactor-grade plutonium, contains a larger proportion of non-fissile isotopes. This increases the amount of plutonium required to make a warhead.

In order for plutonium in reactor spent fuel to be used to make a nuclear weapon, it must first be separated from the residual uranium, which is mostly non-fissile uranium-238 and fission products. This can be done in a reprocessing plant (see Chapter 2).

Table 6 shows the historical growth of world inventories of commercial and military plutonium in the world. Note that the growth of global plutonium inventories has been particularly rapid since about 1980. This is because a large number of nuclear power plants ordered in the 1970s came on line in that decade. Commercial nuclear power plants now generate in about four years the total quantity of plutonium in military inventories accumulated in the more that five decades since the Manhattan Project reactors were built.

High-Level Waste Management: Short- and Medium-Term Issues

The nuclear industry in the United States has created a perception that there is an urgent need for the U.S. government to take over the management of spent fuel by 1998. While the government made the commitment to do so in the 1982 Nuclear Waste Policy Act, there is no technical justification for this demand.

Table 6: Historical World Plutonium Inventories
(in Metric Tons)

Type	1945	1950	1960	1970	1980	1990	1995
Subtotal: Military	0.1	2	45	130	210	265 (note 6)	270 (note 6)
Unseparated Commercial	0	0	0	1	145	530	805
Separated commercial	0	0	0	5	40	120	195
Subtotal: Commercial	0	0	0	6	185	650	1,000
Total	0.1	2	45	136	395	915	1,270

Sources/Notes:

1. Figures are rounded either to one significant figure or to the nearest five metric tons, as indicated. The total row is not further rounded.
2. No country has released historical military plutonium production data. Only the U.S. has released official military plutonium production data. All other military data are rough estimates.
3. Military plutonium figures for the U.S. were derived from official U.S. DOE data for cumulative production as of 1993 (which is the same as that for 1988 when production stopped). Other figures were estimated by IEER from Cochran et al. 1987; Cochran and Norris 1993, Albright et al. 1993, and von Hippel et al. 1986.
4. Figures for commercial separated and unseparated plutonium for 1980 and 1990 are from Albright et al. 1993. The 1995 commercial data assume an overall discharge of spent fuel of about 70 metric tons per year in the 1990s, of which 15 metric tons per year are separated, the rest remaining unseparated. Figures for commercial separated plutonium for 1970 were estimated from data in Albright et al. 1993.
5. Separated commercial plutonium is owned by the following countries

that are currently reprocessing: France, Britain, Japan, Russia, India. In addition, countries that have no current reprocessing have contracts for reprocessing with France and Britain. These countries also own substantial commercial plutonium stocks. They are Germany, Belgium, Holland, Italy, and Switzerland. The United States also has a relatively small stock of commercial plutonium from its West Valley reprocessing plant, which was shut down in 1972.

6. We have assumed a figure of 150 metric tons of military plutonium for Russia. Recent information from Russia indicates that the figure may be lower, at about 130 metric tons (rounded), or higher. (William Weida, private e-mail, March 25, 1996.)

The government also made many other commitments in that law and related regulations. For instance, the process for the selection of a repository was to be a scientifically sound one. The 1982 law also provides for the creation of two repositories to ensure regional equity. One was to be selected from western and southwestern sites, the other was to be in the east. These commitments have been abandoned in what has become a transparently politicized process, with diminishing scientific content and little environmental merit.[157] In 1987, the U.S. government decided to restrict the study to a single site in Nevada, a politically weak state, on land that is controlled by the U.S. government, but which is claimed by the Western Shoshone people.

Since the official opening date for a repository has been pushed back to 2010 at the earliest, the nuclear industry has demanded that the government take control of the spent fuel anyway by building a Monitored Retrievable Storage (MRS) facility for it. Over the years, many variations of a centralized interim storage have been proposed at many different locations. A relatively recent one involves a consortium of utilities that would build a MRS in a private arrangement with the Goshute tribe (Skull Valley) in Utah.

The nuclear industry argues that it is better to store spent fuel at one centralized MRS than at scores of sites around the country. However, this argument has not been given much technical substance by its proponents. If nuclear power plant operators cannot secure and manage

---

[157] See Pigford 1995; NRC-NAS 1995; and Makhijani 1995 for descriptions of some of the problems and controversies in setting standards for the Yucca Mountain site.

their spent fuel on-site, then it is surely natural to ask whether society should allow them to continue in the far more tricky business of operating nuclear power plants. Further, there has been no technical demonstration that the transportation risks to a centralized site and of storage there are lower than those for on-site storage.[158]

There are other arguments against a MRS:

- Both locations for a MRS facility that have been currently proposed involve Native American land or land claims.

- A "temporary" storage location is likely to become permanent, because once the U.S. government takes charge of it and puts it in one place, there will be strong pressure to leave it there due to the political difficulty of finding a new location to which to move it. The risk that a repository may be constructed at an MRS site at the end of the supposed temporary 40-year storage period is unquantifiable but non-negligible. Since a MRS site is not being selected on the basis of its merit as a repository site, the likelihood of inappropriate disposal under such an eventuality is great.

- The political pressures to reverse U.S. non-proliferation policy and reprocess commercial spent fuel are likely to increase if spent fuel is stored at a single site. The environmental and health risks to workers and the communities near the MRS site would greatly increase should reprocessing actually be carried out.

While there are problems associated with expanding on-site storage at certain locations, they have less to do with on-site storage, as such, than with (i) whether the reactors were wisely sited in the first place, and (ii) the integrity with which storage casks and their foundations are built, inspected, and licensed. In principle, removing fuel from the pools when they have cooled sufficiently and storing it in properly built dry casks that are situated on foundations appropriate for the local geology and environment should provide storage that is comparable in risk per cask to a MRS. Moreover, it could do so without the associated risks of a MRS cited above. These problems of storage

---

[158] MRS Review Commission 1989, pp. 52, 45.

at power plants can and should be settled by open, democratic processes available within the public utility commission jurisdictions that cover those plants and not by creating a policy to build a MRS, which would compound the errors of the past. In order for decisions to be made within such a context, it is essential that the NRC not short-circuit the process by licensing casks generically, thereby preventing local communities and citizens from participating in it.

A part of the urgency expressed by nuclear utilities for a repository or a MRS derives from the fact that many utilities are obliged to spend money to expand on-site storage facilities despite the payments that they have made into the nuclear waste fund. There is some merit to this portion of this utilities' argument. Because their money has been wasted and the government cannot reasonably fulfill its commitment to move the spent fuel from their sites by 1998, they are having to spend extra money for expanding on-site storage. But this problem cannot be used to increase risks by moving fuel from one place to another and avoid the responsibilities of putting in place a long-term management approach that will work far better than the dismal program now in place. Present policy needs to be replaced by a straightforward, three-part policy:

- allocating a portion of the nuclear waste fund to expanding some on-site storage facilities or storage facilities as close to the point of origin as possible if the reactor site is found especially unsuitable for some reason;

- abandoning the present flawed repository program;

- focusing efforts on restructuring the long-term waste management program so that it acquires the needed level of technical and environmental merit.

The nuclear industry has not chosen to pursue such a course because the argument about waste management in the coming years and decades is, in large measure, not about sound waste management but about the future of nuclear power.

1. Nuclear Waste and Nuclear Power

For nuclear power proponents, the "solution" to the nuclear waste problem seems to be defined primarily by the criterion that the waste should be removed from power plant sites. This nuclear waste shell game, of course, does nothing to solve the problem for society even in the present, to say nothing of future generations. But it does address one concern of the nuclear establishment that arose out of a 1978 interpretation of California law that led to a moritorium on the construction of new nuclear power plants until the state was satisfied that a means of permanently disposing of high-level nuclear waste had been approved by a federal agency with the authority to do so. The law was upheld by the California Supreme Court in 1983.[159] Other states have followed California's lead with similar laws. This decision connected the future of nuclear power to the nuclear establishment's ability to show the problem had been "solved." But it shifted the definition of what a solution meant to the industry from environmental, health, and non-proliferation criteria to simply transferring the private liability of spent fuel to the federal government.

The nuclear industry's urgency for a "solution" to the nuclear waste problem stems in large part from the belief that if it demonstrates that it can pass off the liability to the government, then new nuclear power plant orders may resume. The pressure for an unrealistically early date for repository opening in the 1982 Nuclear Waste Policy Act stemmed from this assumption. This sense of urgency was abetted by those concerned about arms control. An early repository opening date meant that spent fuel could not be reprocessed for its plutonium content. It would put the seal on U. S. policy against reprocessing that had been initiated under President Ford in 1976 and strengthened by President Carter.

Opponents of nuclear power saw in the design for a quick repository opening two central problems. First it would make an environmentally sound search difficult or impossible. Second, it would remove a roadblock to new nuclear power plant orders.

The connection between the future of nuclear power and problem of managing some 30,000 metric tons of spent fuel that already exist is not a necessary one. The nuclear power-nuclear waste issue can be more easily understood by noting that there are three different kinds of spent fuel society must consider in connection with a long-term waste management program:

---

[159] Carter 1987, pp. 84-89.

- Over 30,000 metric tons of spent fuel already exist from past nuclear electricity production. This must be managed no matter what the future of nuclear power.

- A certain quantity of additional spent fuel, most likely several tens of thousands of metric tons, will be generated as a result of further operation of existing nuclear power plants, since nuclear electricity now constitutes too large a share of the country's electricity supply to be switched off suddenly whatever one's position on the desirability or lack thereof of nuclear energy (see Chapter 9).

- Whether any spent fuel from future nuclear power plants will be generated depends on future nuclear power plant orders in the United States—and none are on the horizon. These wastes do not have to be created at all.

While the U.S. government assured existing reactor owners and operators that it would take responsibility for long-term management and disposal of spent fuel, it undertook no such obligation for future nuclear power plants. Since no new plants have been ordered since 1978, the break between existing plants and future plants is a clean one and the U.S. government could refuse to accept liability for the latter. The area of controversy can then be reduced to the wastes to be generated by further operation of existing nuclear power plants—that is, how long existing plants will operate and whether nuclear power plants can be phased out consistent with other goals such as reduction of greenhouse gas emissions from fossil fuel burning (see Chapter 9).

2. *Reprocessing*

The flawed and much delayed repository program is creating new pressures for a return to reprocessing of civilian spent fuel in the United States. Civilian spent fuel was reprocessed in the United States from 1966 to 1972, when the $32 million[161] reprocessing plant at West Valley in western New York, owned and run for the most part

---

[161] A $32 million investment in 1966 would be equivalent to about $160 million in 1995 dollars.

by Getty Oil, was shut, leaving state and federal taxpayers with billions of dollars in waste management liabilities.

Reprocessing of civilian spent fuel, called commercial reprocessing, is being carried out in five countries: France, Russia, Britain, Japan, and India. The stated purpose is to use the plutonium as a reactor fuel. All five reprocessing programs are government-owned or subsidized. This has not prevented calls by some in the nuclear industry and in Congress for a return to reprocessing in the United States, despite stated U.S. government policy to discourage reprocessing on non-proliferation grounds. Cogéma and British Nuclear Fuels Limited (French and British government-owned companies, respectively) are actively seeking business in the United States.

The use of plutonium as a fuel in nuclear reactors after it is reprocessed is also proving to be a major economic problem. Throughout the world, breeder reactor programs are faltering despite four decades of development; a few remain open only due to large government subsidies. The use of plutonium oxide mixed with uranium oxide as a fuel (MOX fuel, for short) in light water reactors has also had significant setbacks due to the high cost of fabricating MOX fuel. Independent studies have found that it is cheaper to use uranium fuel instead of MOX fuel even if the plutonium is free (as it might be if the source were surplus plutonium from weapons programs).[161]

The clearest, most detailed proposal to resume reprocessing and connect the military side of the nuclear establishment to the civilian side in a manner that has not been done in the United States is discussed in an August 1995 report by Westinghouse, which is both a major DOE contractor and a vendor of nuclear reactors. This report called for the DOE to start accepting civilian spent fuel by 1998 and reprocess it at Savannah River Site (where Westinghouse is the prime contractor).[162]

Reprocessing civilian spent fuel would be a very costly enterprise. In fact, the high cost of reprocessing is one of the main reasons that makes recycled plutonium an uneconomic fuel. A 1996 National Research Council report estimates that costs of reprocessing in a new plant in the U.S. would range from $810 to $2,110 per kilogram of heavy metal in spent fuel (in 1992 dollars), depending on whether it is

---

[161] NAS 1995.
[162] Westinghouse 1995.

governmental or private enterprise, respectively. The costs include vitrification of liquid high-level wastes resulting from reprocessing, and are based on European experience. These are minimal estimates, based on actual operating data, and do not include costs for high efficiency removal of transuranic elements for the purpose of waste disposal.[163]

Even at the lower end of the above range of costs, the overall bill for reprocessing about 80,000 metric tons of spent fuel that is projected to be generated by U.S. nuclear power plants during the licensed lifetimes would be about $65 billion. In addition, there would be tens of billions of dollars of costs for repository disposal and MOX fuel fabrication. We estimate each of these to be on the order of $25 billion in 1992 dollars, so that the overall costs of this approach of spent fuel management and disposal would be on the order of $115 billion (1992 dollars) or almost $130 billion in 1995 dollars.[164]

If the reprocessing is done at commercial rates, the costs of unsubsidized reprocessing, vitrification, and long-term waste management would be about 0.9 cents per kWhe, in 1992 dollars or about one cent in 1995 dollars. This is ten times the amount that nuclear utilities are now required to pay into the Nuclear Waste Fund. The overall costs for 80,000 metric tons of spent fuel would be on the order of $240 billion

---

[163] NRC-NAS 1996, pp. 112-117. The government-owned plants are cheaper because they are subsidized. For instance, the costs of capital are assumed to be far lower; taxes and return on investment are assumed to be zero. (See Tables 6-8, p. 115, and Appendix J.

[164] The NAS has estimated that using MOX fuel would result in an incremental fuel cost of 0.3 cents/kWhe, including tax and insurance (NAS 1995, p. 302). There is not much difference between costs using commercial MOX experience and those estimated for MOX derived from weapon-grade plutonium. Assuming that a reactor is fueled with one-third MOX fuel, the incremental cost per kilowatt hour overall of using MOX fuel would be about 0.1 cent/kWhe. This amounts to $310 per kilogram of spent fuel having a burn-up of 40,000 megawatt-days-thermal/metric ton of heavy metal, yielding a total cost of about $25 billion for 80,000 metric tons of fuel in 1992 dollars. The cost of repository disposal has been estimated in 1988 dollars at about $265,800 per metric ton of spent fuel (see Makhijani and Saleska 1992, pp. 66-68). Assuming no real cost escalation since that time, this amounts to $330,000 in 1995 dollars or about 0.11 cents per kWhe, for 40,000 MWdth/metric ton burn-up. The added cost for 80,000 metric tons of spent fuel for the same burn-up is about $26 billion in 1995 dollars. If the plutonium extracted by reprocessing is vitrified with fission products, the cost is estimated to be of the same order of magnitude as MOX fuel fabrication.

(1995 dollars).[165]

Reprocessing civilian spent fuel would also greatly exacerbate a high-level waste management problems in the U.S. The volume of liquid high-level radioactive wastes in the U.S. from military programs is an order of magnitude larger than all civilian spent fuel (see Table 4), even though the amount of heat generated from fission was far lower. This is because reprocessing wastes in the U.S. have been neutralized, which greatly increases their volume. Current tanks at Savannah River are made of carbon steel and therefore cannot accept acidic wastes. Therefore, high waste volume due to civilian spent fuel would increase greatly if the Westinghouse proposal is implemented. The prospects for the vitrification of most existing waste are uncertain due to continuing technical problems with the processing of waste for vitrification.

The reprocessing plants at Savannah River Site are over 40 years old. In 1992, DOE suspended its program of upgrading the plants, saying they would be shut down after limited runs. But the DOE decided in late 1995 to keep them open for a number of years, under the guise of "environmental management" to reprocess its own spent fuel, some of which is deteriorating in underwater storage pools. But a report by Noah Sachs of the Institute for Energy and Environmental Research showed that this policy would exacerbate environmental problems, increase risks of fires or explosion in high-level waste tanks, and undermine U.S. non-proliferation policy.[166]

The process leading up to the DOE decision to keep one of its two Savannah River Site reprocessing plants open (and the other on standby) apparently gave its site contractor, Westinghouse, the opening

---

[165] Assuming a burn-up of 40,000 megawatt-days thermal per metric ton (which is higher than the average for spent fuel now in the inventory), each kilogram of heavy metal yields an electrical energy output of about 310,000 kilowatt hours. At about $2,100 per kilogram of heavy metal for reprocessing, the cost would be about 0.7 cents per kilowatt-hour electrical, excluding MOX fuel fabrication costs and repository disposal costs. If we make the assumption that the combined costs for MOX fuel fabrication and repository disposal would amount to $50 billion for 80,000 metric tons of spent fuel, the additional cost of these two items works out to 0.2 cents kilowatt-hour electrical. Hence, the combined cost estimate is on the order of 0.9 cents per kilowatt-hour electrical, in 1992 dollars. Note that some recycled fuel will be MOX fuel, which reprocessing technology does not yet handle. The costs of handling this fuel would be higher, but are at present difficult to estimate due to lack of experience.

[166] Sachs 1995.

stunning setback for U.S. non-proliferation policy—one of the goals of which is to discourage reprocessing, and pose serious environmental threats to the people of the region by increasing liquid high-level wastes to be stored in the tanks at the site. Yet the proposal is presented as an environmental program: reprocessing is not even called by that name, but rather "chemical stabilization." This is a rather odd way to characterize the conversion of solid, stable oxides in spent fuel into liquid, chemically reactive nitrates.

It is difficult to overemphasize the central importance of U.S. policy against commercial reprocessing. Since 1976, the United States has been the only nuclear power to renounce commercial reprocessing on non-proliferation grounds. While President Reagan lifted the formal ban against commercial reprocessing, its exorbitant costs have effectively kept the ban in place and hence available for implementation of U.S. non-proliferation policy on plutonium. It is crucial to note that commercial reprocessing is now the most important contributor by far to the growth of nuclear weapons-usable materials in the world. Further, the United States is also the only nuclear weapons state to have announced an end to military reprocessing. The United States is therefore the only country with the political standing to persuade Russia and other countries to stop civilian reprocessing.[167] The United States took one step toward a reversal of that policy in December 1995, when it decided to keep one reprocessing plant at Savannah River Site open in the name of "environmental management." An implementation of the Westinghouse proposal would be a grave practical setback to the implementation of the U.S. non-proliferation policy of discouraging reprocessing abroad.

A strong non-proliferation policy on plutonium is needed urgently because of surplus military plutonium stocks and the growing amounts of commercial separated plutonium in Russia. Plutonium is regarded as a national treasure by the Russian nuclear establishment. Without U.S. prodding, reciprocity, and incentives, Russian plutonium stocks are likely to continue to grow along with potential for diversion into black markets. About 30 metric tons of separated commercial plutonium are stored at the Chelyabinsk-65 plant in the Urals in about 12,000 thermos flask-size bottles. The bottles are suspended in con-

---

[167] For further discussion of this issue, see Makhijani and Makhijani 1995; and NAS 1994.

crete trenches in the floor of the storage building. Safeguarding and verifying these stock is a major non-proliferation issue.

Long-Term Management Issues

The hasty, highly politicized repository program that followed the enactment of the Nuclear Waste Policy Act in 1982 is summarized below:[168]

1982:   The Nuclear Waste Policy Act is enacted into law. A western site to be selected from among those that the DOE was already studying.

1982-   DOE narrows the search for a western site to three places:
1986:   Hanford, Washington; Yucca Mountain, Nevada; Deaf Smith County, Texas. The government controls the land at the former two sites. Critics charge a flawed process for site ranking.

1983:   A National Research Council panel (the Waste Isolation Systems Panel) issues a comprehensive report, funded by DOE.[168] Among other things, the report indicates that radiation dose to a maximally exposed individual from a Yucca Mountain repository could be very high and that Hanford faced serious geological difficulties due to high rock stresses that caused the rock to fracture. The DOE essentially ignores these findings.

1986:   DOE publishes its "Draft Area Recommendation Report," specifying dozens of sites for investigation in politically influential eastern and mid-western states. The DOE chooses not to list for investigation the type of eastern site especially recommended by the 1983 NRC-NAS report. A storm of protest breaks out for a variety of reasons. DOE suspends its eastern repository effort.

1987:   Congress mandates that the DOE restrict its site characterization program to one site: Yucca Mountain, Nevada. Congress and the DOE ignore the 1983 NRC-NAS report (cited above) that indicated high doses to the maximally exposed individual from a repository at this site. The claim of the Western Shoshones to the land is not given consideration.

---

[168] NRC-NAS 1983.

1989: DOE announces that a repository will not open until the year 2010, a 12-year postponement of the original goal.

1989- A concern grows that emissions of carbon-14 from a Yucca
1993: Mountain repository may exceed EPA regulations for high-level waste repositories. The EPA Science Advisory Board finds (by 1993) that Yucca Mountain may not meet the standard. The consequences of carbon-14 emissions for individuals would be minute increases in radiation dose, but the population dose over thousands of years could be high.[169]

1990s: DOE contractors calculations indicate that use of groundwater contaminated by a Yucca Mountain repository could result in high radiation doses, far above allowable limits.[170]

1992: Congress passes the Energy Policy Act, which mandates that EPA standards generally applicable to high-level waste disposal not be applied to Yucca Mountain. The law requests the National Academy of Sciences to form a committee to advise the EPA on the technical bases for setting standards for high-level waste disposal at the Yucca Mountain site.

1993- DOE contractors calculate that doses from a Yucca Mountain
1994: repository to a future subsistence farmer would greatly exceed prevailing EPA standards, echoing qualitatively earlier work by the National Research Council of the National Academy of Sciences.

1995: The National Research Council of the NAS issues its report, *Technical Bases for Yucca Mountain Standards*, but it is not unanimous. The majority recommends a new, untried method for dose calculations that could have the effect of reducing calculated dose. One panel member, Professor T.H. Pigford ofthe University of California at Berkeley, files a vigorous dissent. He claims that the committee majority had engaged in creating "non-scientific policy fixes" to reduce calculated doses, that the method for calculating doses was not mathematically valid, and that it would result in an intolerable

---

[169] EPA 1993.

[170] For various dose estimates from a Yucca Mountain repository, see "Centerfold for Technoweenies," *Science for Democratic Action*, Institute for Energy and Environmental Research, Vol. 4. No. 4, Fall 1995, pp. 8-9.

loosening of generally accepted radiation protection standards.[171]

1995-   Congress considers several nuclear waste bills. Provisions
1996:   range from the appointment of a presidential commission on
             nuclear waste, to relaxation of radiation exposure standards,
             to mandating the disposal of waste in a Yucca Mountain
             repository, and to forcing DOE to start taking possession of
             the waste in 1998 and moving it to a MRS at Yucca Mountain.
             Congressional sentiment in favor of reprocessing grows.

Concluding Observations

The state of knowledge of long-term problems on both the security and environmental fronts is today insufficient for society to decide on long-term disposal of spent fuel. Alvin Weinberg, the first director of Oak Ridge National Laboratory, opined in 1972 that the public ought to trust the nuclear priesthood that was responsible for guarding nuclear weapons with managing nuclear waste.[172] The revelations of the past two decades, such as systematic environmental mismanagement, fabricated data, cover-ups, and human experiments without informed consent, have eroded any faith in that priesthood that the public may have had. In the meantime, reliance on nuclear power has grown and the already large quantities of weapons-usable plutonium in the world are rising rapidly.

The story of the volume of wastes is quite different. By far the largest volume of waste created is in the uranium mine wastes and uranium mill tailings, which are disproportionately on Native American lands.[173] Depleted uranium, uranium-processing waste, and "low-level" wastes from routine operations constitute most of the volume. Table 4 (page 111) shows the volume and radioactivity content of wastes generated in the U.S.

---

[171] NRC-NAS 1995. For Professor Pigford's dissent, see Appendix E of NRC-NAS 1995. Also see Pigford 1995; and Makhijani 1995 for further discussion.

[172] Weinberg 1972.

[173] Tribal and colonial peoples have borne a disproportionate amount of the negative environmental impact from uranium mining and milling throughout the world. See Makhijani et al., eds. 1995, Chapter 5.

Yucca Mountain. View of the north portal entrance to the Exploratory Studies Facility just after the Tunnel Boring machine moved into the 200-foot starter tunnel. Nuclear industry pressure to find a "solution" for spent fuel and other high-level waste has led the government to impose an unrealistically hurried schedule for a repository opening, despite a number of questions about the suitability of this site. (U.S. Department of Energy)

PART TWO:
MORE NUCLEAR POWER PLANTS
OR A SOUND ENERGY POLICY?

It will not be possible to provide the energy needed to bring a decent standard of living to the world's poor or to sustain the economic well-being of the industrialized countries in environmentally acceptable ways, if the present energy course continues. The path to a sustainable society requires more efficient energy use and a shift to a variety of renewable energy sources.

—Johansson et al., eds. 1993[173]

---

[173]Johansson et al., eds. 1993, p. 9.

CHAPTER 7:
"INHERENTLY SAFE" REACTORS—
COMMERCIAL NUCLEAR POWER'S
SECOND GENERATION?

Nuclear power has been promoted in the late 1980s and early 1990s in a number of ways, in the attempt to regain ground lost in the wake of the 1979 partial meltdown in Unit 2 of the Three Mile Island nuclear plant in Pennsylvania. This has included diverse efforts, such as DOE and nuclear industry intervention on behalf of the beleaguered Shoreham and Seabrook nuclear plants in New York and New Hampshire (the latter opened, the former did not), to major stories on the cover of *Time* magazine.[174] A wide array of nuclear industry groups, operating in a coalition named the Nuclear Power Oversight Committee, had the ambition of seeing a new commercial nuclear plant ordered by the mid-1990s.[175] As another example, the Nuclear Energy Institute in 1995 published the fifth annual update of its plan to revive nuclear power.[176]

This time around the industry has also been putting forward an environmental rationale as part of its promotion of nuclear power. Its spokespersons state that nuclear power could be a principal factor in cutting back emissions of pollutants, notably carbon dioxide which results from the burning of fossil fuels. As an article in *Fortune* magazine put it, "concern about the earth's rising temperature could turn a technological pariah into a savior—if new reactor designs overcome

---

[174]Greenwald 1991.

[175]NPOC 1990. The Nuclear Power Oversight Committee includes representatives of nuclear utilities (Commonwealth Edison Company and Duke Power Company), nuclear industry groups (EPRI, EEI, ANEC, INPO, U.S. CEA, and NUMARC), the major nuclear vendors GE and Westinghouse, as well as power plant construction firms (Bechtel, ABB, and Combustion Engineering).

[176]NEI 1995.

worries about atomic safety."[177]

To address the safety concern, the nuclear industry has been promoting a "second generation" of commercial nuclear power reactors, some of which have been labeled "inherently safe" by their proponents. Is that claim reasonable? Another big question is whether such plants can be economically viable. And is nuclear power the appropriate way to address concerns about the build-up of greenhouse gases?

The safety question is a central one, since public skepticism of industry claims grew greatly after the Three Mile Island and Chernobyl accidents. But the road will be a hard one, perhaps impossible due to the choices that were made in the initial development of nuclear power. The safety issues surrounding nuclear power are especially difficult not only because of the technological complexity of the plants, but also because of the potentially catastrophic and irreversible consequences of severe accidents. For instance, after the accident at Three Mile Island, one of the investigators made the following comment:

> If there is one thing that I have learned through the [Three Mile Island] investigation, it is this: Nuclear power plants are very large, very complex systems that cannot be completely accurately modeled. Dangerous transients cannot be incurred deliberately so that the actual plant response to all events can be experienced and tested. . . . Current plant performance statistics must not be accepted as "good enough" because they may not be good enough for the future, and one accident is one too many.[178]

The NRC also re-evaluated its position on the safety of LWRs; there was a general realization that despite all the studies and analyses that had been done, nuclear plants were not nearly as safe as had been assumed. In the mid-1970s, some believed that the probability of an accident at an LWR involving severe core damage was on the order of one in one million per year of reactor operation. The experience of the Three Mile Island accident, along with subsequent plant-specific probabilistic risk assessments in the mid-1980s led to a revision. The reassessment indicated that, on average, the likelihood of a severe accident at existing reactors may be closer to one in 3,000 per year of reactor

---

[177]Faltermayer 1988, p. 105.
[178]Jaffe 1981, p. 45.

operation, or about 300 times more likely than previously thought.[179]

NRC Commissioner James Asselstine, in Congressional testimony in 1986, put the results of the NRC reassessment this way:

> The bottom line is that given the present level of safety being achieved by the operating nuclear power plants in this country, we can expect to see a core meltdown accident within the next 20 years; and it is possible that such an accident could result in off-site releases of radiation which are as large as, or larger than, the releases estimated to have occurred at Chernobyl.[180]

The safety problems with the current generation of reactors have contributed to the widespread resistance to nuclear power. As even the NRC has acknowledged,

> Public acceptance, and hence investor acceptance, of nuclear technology is dependent on demonstrable progress in safety performance, including the reduction in frequency of accident precursor events as well as a diminished controversy among experts as to the adequacy of nuclear safety technology.[181]

Advocates for a second generation of nuclear plants imply or state that this safety problem has been solved with new designs that rely on "passive" or "inherent" safety features. We will discuss some of these designs in the section below.

New Reactor Designs: An Overview

Although there is strong support among nuclear industry manufacturers for the idea of a revitalization of nuclear power, there is considerable debate as to what type of reactor technology should be used for the job.

Many designs have been proposed, ranging from those that are only modest modifications of current light water designs, to substantially different designs cooled by gas or liquid sodium metal. The basic concepts underlying these latter designs have been around as long as

---

[179]UCS 1990, pp. 1-6, 1-7.

[180]Asselstine 1986.

[181]U.S. Nuclear Regulatory Commission, "Policy Statement on Severe Reactor Accidents Regarding Future Designs and Existing Plants," *Federal Register,* Vol. 50, p. 32138 (August 8, 1985), as cited in UCS 1990, p. 1-3.

nuclear power, but they were edged out of the market in the early years by the light water reactor bandwagon. The "advanced" reactors tend to have the following features;

- lower overall thermal power output;

- smaller thermal power density in the reactor;

- greater reliance on "passive" safety systems (i.e., systems which do not depend on the actuation of engineered machinery, but rather on natural physical processes and forces, such as gravity).

There is a basic split between the "old guard" which continues to advocate the basic light water design, with improvements, and those who back non-light water designs in the hope that their design will alleviate the widespread concerns about safety associated with light water reactors. Proponents of the LWR believe that its safety can be improved and that investment in its continued dominance of the market is the best strategy for promoting nuclear power. Advocates of newer designs believe that nuclear power may gain more public acceptance through use of the term "inherently safe." But while these designs have been given new life under the label "inherently safe," it has raised concerns among the light water old guard that this nomenclature implicitly brands the LWR as inherently *un*safe.

The promoters of variations on the light water design tend to be those who have strong economic interest in reinvigorating the industry on the basis of the existing light water technology, in which they have substantial investment. For instance, the Nuclear Power Oversight Committee, mentioned above, has stated that

> The extensive operating experience with today's light water re-
> actors (LWRs), and the promise shown in recent technical de-
> velopments, leads the industry to the conclusion that the next
> nuclear plants ordered in the United States will be advanced
> light water reactors (ALWRs).[182]

According to this industry group, the advantage of the light water reactor approach is that it "rel[ies] on proven technology."[183] This

---

[182] NPOC 1990, p. I-2.
[183] NPOC 1990, p. I-2.

attitude is shared by similarly situated industry members in Europe. Karlheinz Orth, an official of the nuclear division of Siemens AG, for example, has said that the contrast between the effects of the Three Mile Island and Chernobyl accidents proved the basic soundness of the pressurized light water reactor. But at the same time, Orth criticizes some of the new designs for an over-reliance on passive systems. As he told an international safety conference in 1988:

> The importance of passivity is overestimated. Every reactor concept is based on certain inherent safety features and also depends on active and passive engineered safety features. . . . Where reliance is placed solely on inherent safety features or on purely passive engineered safety features, it would not be possible for an operator to select or even influence the final condition of the plant. . . . There is no reason to leave today's mature LWR technology only in order to experiment with . . . half-developed but "alternative" concepts. Preferences established by publicity can be no substitute for operational experience.[184]

To advocates of advanced reactor designs, it is precisely such attitudes that are hurting the chance for the development of a new generation of safer reactors. In the words of an advocate of a helium-cooled, carbon-moderated reactor called the MHTGR (Modular High-Temperature Gas-Cooled Reactor):

> Deployment of a qualitatively different second generation of nuclear reactors can have important benefits for the United States. Surprisingly, it may well be the "nuclear establishment" itself, with enormous investments of money and pride in the existing nuclear systems, that rejects second-generation reactors. It may be that we will not have a second generation of reactors until the first generation of nuclear engineers and nuclear power advocates has retired.[185]

We briefly review some of the proposed light water designs and then do the same for the MHTGR.

---

[184]As quoted in *Nucleonics Week* 1989, p. 10.
[185]Lidsky 1988, p. 226.

New Light Water Reactor Designs

The range of new light water designs has been divided into two broad classes: *evolutionary* light water designs (which are similar to recent LWRs, but have enhanced features), and *advanced* water-cooled designs (which, although water-cooled, have designs which are significantly different than recent LWRs).[186] Several examples of evolutionary LWRs include:[187]

- the General Electric Advanced Boiling Water Reactor (ABWR), a few of which have already been ordered in Japan and one of which has been completed; [188]

- the Westinghouse SP-90 advanced PWR;

- Combustion Engineering's System 80+ PWR.

Evolutionary light water reactors tend to be large (the above models are 1,100 MWe or larger). They incorporate some improvements in fuel-cycle efficiency and safety features, but are not fundamentally different from current designs.[189] Thus, they have the same essential safety weakness of current reactors: the risk that the reactor core could melt down in the event of a loss-of-coolant accident and cause a catastrophic release of radioactivity. Being large, they also are more financially risky for utilities due to the unpredictability of growth in demand for electric power. G.E.'s ABWR design and Combustion Engineering's System 80+ received approval for their final designs from the NRC in 1994, and final design certification in 1997.[190]

Advanced light water reactor designs, on the other hand, tend to be smaller than the evolutionary designs (ranging from 320 to 750 MWe), incorporating many features that are referred to as "passive" safety systems. They also tend to incorporate modular construction features which it is claimed will reduce the costs of the plants. Some

---

[186]UCS 1990, p. 2-11.

[187]As listed in UCS 1990, pp. 2-11 and 2-12.

[188]In late January 1996, the first ABWR was commissioned and connected to the electricity grid in Japan. See *Nuclear Energy Insight96*, Nuclear Energy Institute, Washington, D.C., February 1996, p. 1.

[189]UCS 1990, p. 2-12.

[190]NEI 1998.

examples of advanced light water designs are:[191]

- the Advanced Passive PWR (AP-600) which is being developed by Westinghouse;

- General Electric's Simplified Boiling Water Reactor (SBWR);

- Asea Brown Boveri Atom's "Process Inherent Ultimate Safety" (PIUS) reactor, in both pressurized water and boiling water version;

- the Safe Integral Reactor (SIR) pressurized water reactor, developed by a team led by Combustion Engineering.

Much is being made by industry of the "passive" or "inherent" safety features of these light water designs, such as the "emergency cooling features, which depend more on natural processes such as gravity than on powered equipment such as pumps."[192] They also tend to include a simplification of the overall design. For example, Westinghouse's AP-600 pressurized water reactor requires 50 percent fewer large pumps and heat exchangers, 60 percent fewer valves and pipes, and 80 percent less control cable.[193] The power density in its core is lower as well, at 74 kW per liter in comparison to 108 kW per liter for a conventional Westinghouse PWR.[194] The industry backers of these designs hope that such simplifications, combined with modular plant construction and streamlined NRC licensing, will lower construction costs and schedules. Some of these features might reduce the risk of serious accidents, but not eliminate it. The potential for loss-of-coolant, runaway chemical reactions between fuel cladding and steam, and catastrophic meltdown will apparently remain. The G.E. and Combustion Engineering advanced reactor designs received design certification from the NRC in 1997.[195]

Other Reactor Designs

So-called "inherently safe" reactor designs include those which are

---

[191]UCS 1990, p. 2-12.
[192]NPOC 1990, p. I-2.
[193]*Nucleonics Week* 1989, pp. 3-4.
[194]*Nucleonics Week* 1989, p. 6.
[195]NEI 1998.

substantially different from the light water designs referred to above. These include several new versions of liquid metal-cooled breeder reactors, such as the General Electric Power Reactor Inherently Safe Module (PRISM) design and Rockwell International's Sodium Advanced Fast Reactor (SAFR). As with other supposedly "inherently safe" designs, these claims appear to be more in the realm of propaganda than fact. Any reactor which uses liquid sodium as a coolant is vulnerable to the violent chemical reactions and potential explosions which can occur from contact of sodium metal with water. Moreover, use of breeder reactors generally involves reprocessing of spent nuclear fuel and plutonium fabrication activities, which bring with them a whole host of safety, proliferation, and environmental issues.

1. The MHTGR

Much of the attention regarding "inherently safe" reactors has been garnered by the modular high-temperature gas-cooled reactor (MHTGR). As noted above, this design relies on helium for coolant and is graphite moderated. As is the case with the pressurized light water reactor, the coolant circulates through steam generators which turn water to steam and drive a steam turbine.[196]

Although there is a fair amount of experience with gas-cooled reactors in Britain (the Magnox design), there have been only two operating commercial plants in the U.S. that have used the high temperature gas-cooled design: Peach Bottom-1, a 40 MWe reactor in Pennsylvania, which operated from January 1967 to November 1974, and the 330 MWe Fort St. Vrain reactor near Denver, Colorado. Neither of these plants is still operating.

It is noteworthy that the Fort St. Vrain HTGR in the U.S. had a lifetime capacity factor of 14.5 percent, an availability factor of 30.9 percent, and a forced outage rate of 60.8 percent before it was permanently shut down in August 1989, partly due to its uneconomic performance.[197] In fact, by the above measure, *the Ft. St. Vrain HTGR was the single, worst-performing commercial reactor in the U.S. nuclear*

---

[196] It has also been proposed that an MHTGR design could operate using a direct-cycle gas turbine instead of a steam generator system. With a direct-cycle gas turbine, the hot helium gas from the reactor core directly drives a turbine, with no water-to-steam cycle involved at all (Lidsky 1986).

[197] UCS 1990, p. 5-8, citing NRC's *Gray Book* (NUREG-0020).

industry.

The variation of this design most generally discussed in the current debate is of the "modular" variety, so named because each reactor unit is considerably smaller (100 or 150 MWe) than many of the existing reactor units, so that a plant would be made up of several "modules." The smaller versions of the HTGR are claimed to be meltdown proof.

Because it has received so much attention among advanced reactor designs, and because it is one of the electricity-producing designs that may be used for military plutonium disposition and tritium production (although DOE is not now actively considering it), we will discuss this design in some detail.

Background

Although there are several variations, the basic MHTGR design in the U.S. is being promoted by General Atomics. Their promotional literature describes the reactor in the following words:

> The MHTGR is a second generation nuclear power system which can satisfy the concerns of the public, the government, the utilities and the investor community about nuclear safety and investment protection. Based on technology developed and demonstrated in the U.S. and Germany, the unique system makes use of refractory coated nuclear fuel, helium gas as an inert coolant and graphite as a stable core structural material. The safety and protection of the plant investment is provided by inherent and passive features not dependent on operator actions or the activation of engineered systems. The high performance MHTGR provides flexibility in power output and siting, competitive energy costs, and can serve diverse energy needs both domestically and internationally.[198]

According to General Atomics, the basic safety idea which the MHTGR, its different fuel design, combined with a size limitation for the reactor (hence the term "modular"), is supposed to make one of the most commonly feared accident scenarios—the core meltdown—impossible.

In contrast to the zircaloy metal-clad uranium oxide ceramic pel-

---

[198]GA (undated), p. 1.

lets of a light water reactor, the MHTGR fuel form is designed to withstand a much higher temperature. Instead of being arranged in vertical rods, the fuel is in the form of millions of tiny spheres, each about the size of a grain of sand. The fuel "kernels" of these spheres (about 350 microns in diameter) consist of enriched uranium mixture of uranium oxide and uranium carbide. The fuel kernels are coated with two layers of pyrolytic carbon and one layer of silicon carbide. Thorium oxide grains for breeding uranium-233 fuel are similarly configured. A full core of fuel is designed to contain a total of about ten billion fuel kernels, which are sealed in vertical holes in graphite blocks. The graphite acts as the neutron moderator.[199]

Normally, during operation or shut-down, the heat generated by MHTGR fuel is carried away by helium gas coolant. If the main heat transfer systems become unavailable through accident or mishap, there is a system which uses natural circulation of air to passively carry away the heat. This system, called the Reactor Cavity Cooling System, operates by natural circulation of outside air through cooling panels along the reactor walls. This system does not depend on active components like pumps, or on actions taken by operators. If by some means even this system is disabled (by vent blockage, for example), the reactor's proponents claim that direct heat conduction from reactor vessel to the reactor cavity wall to the ground is sufficient to remove decay heat without resulting in significant releases of radioactivity from failed fuel elements.[200]

MIT nuclear engineering professor Lawrence Lidsky, an advocate of another variation of the MHTGR, has described the safety features of the basic design:

> The . . . radically different fuel form . . . is capable of withstanding very high temperatures. The [MHTGR] reactors are small to ensure that it is physically impossible for such temperatures ever to be achieved. Such reactors are termed "inherently safe." They are sometimes labeled "passively safe" because no action whatever need be taken to mitigate the effects of equipment failure. Whatever the name, these new reactors eliminate the need for the defense-in-depth strategy. They are designed so that the power plant could suffer the simultaneous failure of all

---

[199]UCS 1990, pp. 2-28 and 2-29.
[200]As summarized in UCS 1990, p. 2-33.

its control and cooling systems without any danger to the public living near the power plant.[201]

The claims of "inherent safety" for the MHTGR are based on its ability to withstand a loss-of-coolant accident without a catastrophic release of radioactivity. The power density and overall reactor size are substantially smaller in the MHTGR relative to present light water reactors, while at the same time the temperature at which its fuel fails is higher than the zircaloy cladding of LWR fuel.[202] But this does not mean that the reactor cannot suffer a loss-of-coolant-accident even in theory. However, the time-scale over which such an accident might develop would be far longer than with an LWR, and the fuel design would help reduce releases of radioactivity, especially if the reactor design incorporated secondary containment.

Safety Concerns

In considering the safety characteristics of the MHTGR, it is well to recall the warning of a British survey, which commented that advanced reactor designers "tend to concentrate . . . on one particular aspect such as a [loss-of-coolant accident], and replace all the systems for dealing with that with passive ones. In so doing, they ignore other known transients or transients possibly novel to their design."

In this context, it is useful to note that the principal original safety concern when nuclear reactor technology was under development was not that they might melt down, but that they might explode due to heating caused by a runaway nuclear reaction. This could result from an inadvertent increase in the multiplication factor causing the reactor to become supercritical (see Chapter 2 above). Neutrons are what cause the fission reaction and, in some cases, the neutron spike accompanying a sudden supercriticality can lead to an explosion of the reactor core. It is this sort of event that occurred at the Chernobyl reactor unit four in the Soviet Union on April 26, 1986, resulting in a catastrophic release of fission products to the environment (see below).

Such a concern was also present in the early days of U.S. nuclear power, particularly with regard to the proposed use of liquid metal

---

[201]Lidsky 1987.
[202]UCS 1990.

cooled fast breeder reactors, such as the Fermi-I reactor which was built near Detroit, Michigan, before a partial meltdown in 1966 damaged its reactor core.

In an ironic historical footnote that carries an important cautionary lesson for the current debate, it is interesting that the term "inherently safe" appears to have first been applied to the *light water* reactor precisely because its design was resistant to large positive reactivity insertions which could lead to a runaway power excursion accident. For example, a 1955 *Popular Science* magazine article lauded the Indian Point-1 reactor then under construction near New York City because of its use of the "Old reliable PWR" design, which was characterized by the article as "inherently safe" because of its "built-in gentleness."[203]

The MHTGR design, it is interesting to note, *is* apparently susceptible to large reactivity insertion events. As stated in the Union of Concerned Scientists' analysis of advanced reactors:

> . . . we do not consider it to be *"inherently safe"* that the MHTGR design experiences a very large reactivity insertion if a control rod ejection accident should occur. In the case of a control rod ejection, the reactor coolant system boundary is breached, and a large reactivity insertion (combined with access of the coolant and/or core to the atmosphere) could result in a very large release of radioactivity to the environment.[204]

A Nuclear Regulatory Commission study of the MHTGR stated, "Both DOE and [Oak Ridge National Laboratory] calculate that the rapid ejection of a control rod could cause the reactor to go prompt critical. For this reason, the potential for rod ejection from the MHTGR must be precluded by design as it is for Fort St. Vrain."[205]

In the Fort St. Vrain reactor (which, as mentioned above, is one of only two power-generating HTGRs that have actually operated in the U.S.) the control rod ejection issue was addressed by two redundant structural systems designed to prevent such ejection. However, as the UCS advanced reactor study notes: "This feature of Fort St. Vrain is

---

[203]Mann 1955.

[204]UCS 1990, p. 3-5.

[205]P.M. Williams, T.L. King, and J.N. Wilson, *Draft Preapplication Safety Evaluation Report for the Modular High-Temperature Gas-Cooled Reactor,* NUREG-1338, NRC Office of Nuclear Regulatory Research, March 1989, p. 4-21, as cited in UCS 1990, p. 3-5.

an engineered safety system solution to an important safety issue; it does not represent an *'inherently safe'* design."[206]

In addition to the potential for reactivity insertions, several other potential safety concerns associated with the MHTGR were discussed in the Union of Concerned Scientists' advanced reactor study. These include:

- *Water contamination of reactor core*: Several events, such as the failure of a steam generator tube or shutdown cooling system, can result in water entering into the reactor core. This can happen since the helium gas, which circulates through the core and the steam generators, is at a lower pressure than the water which is heated by the steam generators. Thus, a breach of steam generator could lead to water ingress to the normally dry core area. The NRC's Advisory Committee on Reactor Safeguards has also suggested that flooding of the reactor vault (which would be underground) could lead to water entering the reactor core.[207]

  Whatever the mechanism, water ingress to the MHTGR core is another event which leads to power increase due to positive reactivity insertion. Moisture entering the primary system also chemically attacks the graphite core structure and the fuel. Ultimately, chemical attack on the fuel (especially defective fuel elements) combined with elevated temperature could lead to some release of fission products from the fuel. One set of assumptions pertaining to such a scenario results in calculated offsite doses to the thyroid of about 3.8 rem.[208] We note here that leakage of moisture into the helium coolant was a problem at the Fort St. Vrain plant and cooling-system component failures were a cause of poor operation and eventual shutdown of the plant.[209]

---

[206]UCS 1990, fn. 3-11, pp. 3-5 and 3-6.

[207]ACRS letter dated 13 October 1988 from William Kerr (Chair, ACRS) to Lando Zech, (Chair, NRC), "Preapplication Safety Evaluation Report for the MHTGR," at 4, as cited in UCS 1990, p. 3-53.

[208]F.A. Silady, et al., "Safety and Licensing of the MHTGR," *Nuclear Engineering and Design*, Vol. 109, 1988, pp. 278-279, as cited in UCS 1990, p. 3-55.

[209]Pigford 1996.

- *Combustible gases and graphite fires*: Hot graphite reacts with steam to produce carbon monoxide and hydrogen, both of which are combustible gases. (Town gas is produced from coal in this way.) This potential presents itself in the event of steam or water leakage into the normally dry helium-filled core and primary coolant loop. In addition, since graphite—a structural material and moderator present in the MHTGR core in significant quantities—is also flammable, the issue and safety consequences of explosions or fires needs to be thoroughly examined.[210] The British Windscale reactor accident in 1957 and the Chernobyl accident in 1986 both involved graphite fires. Further, a graphite fire in an MHTGR could be far more damaging than the Windscale fire, because the fission products in the MHTGR are contained in the tiny graphite fuel elements that would be on fire. In contrast, the fuel elements and graphite moderator in the Windscale reactor were separate entities, although both were, of course, part of the reactor core.

- *Fire extinguishing system vulnerability*: The NRC has suggested that a water-based fire extinguishing system like that at Fort St. Vrain may be acceptable for the MHTGR as well. How this issue is handled, however, has important safety ramifications. For example, flooding the core with water to extinguish a fire may increase the generation of combustible gases. In addition, as noted above, water in an MHTGR core can result in a positive reactivity insertion and risk of explosion. For example, although a decision was taken to use water to extinguish the British Windscale fire, there was great concern that it might cause an explosion. It is partly for these reasons that the fire following the explosion of the graphite-moderated Chernobyl reactor was extinguished not by water but by dropping dolomite, boron, and other materials into the core by helicopter.[211]

---

[210]UCS 1990, p. 3-52.
[211]UCS 1990, p. 3-52.

- *Potential for sabotage*: Protecting nuclear reactors against deliberate sabotage is generally considered to be a generic safety issue for all plants. In the case of the MHTGR, the highly touted passive heat removal system may also increase the opportunity for and risk from sabotage. This is due to the large ground-level vents upon which this cooling system relies. As the NRC has noted:

> In the advanced reactor designs, air passages of the safety-grade decay heat removal systems provide man-sized passages . . . from the protected area yard to locations where relatively small amounts of explosives in the form of shaped charges could breach the reactor vessel.[212]

It is worth noting that versions of the MHTGR design other than General Atomics' reference design may substantially alleviate some of these concerns. For example, the design, advocated by MIT nuclear engineering Professor Lawrence Lidsky, is only 200 megawatts-thermal in size, and employs a direct cycle gas turbine for generating electricity, rather than the use of a steam generator system assumed in the General Atomics design.[213] The use of a gas turbine would remove the need for using water in the system, other than for cooling the gas before it is recirculated into the reactor. This would greatly reduce concerns having to do with water contamination and the concomitant risks (such as a reactivity insertion, chemical attack on the fuel elements, and generation of combustible gases from reactions with steam). This design variant may be adopted by DOE or General Atomics should an HTGR be built in the United States or in Russia.

The discussion above illustrates how variations on the same basic design can potentially result in significant differences in safety level and operational characteristics. They also indicate that a vigorous and open debate over designs while they are still in the paper and experimental, small-scale stages is likely to result in a better and more economical outcome than making adjustments later on.

---

[212] NRC memo (9 September 1987) from Robert A. Erickson to Thomas L. King, "Advanced Reactor Safeguards Reviews," as cited in UCS 1990, p. 3-46.

[213] Lidsky 1986.

The Semantics of "Inherent Safety"

The general arguments of advanced reactor advocates, some of which may be conceptually plausible and appealing, are difficult to either verify or refute in the abstract. This is because they are all essentially in the design stage, with only very limited details made public. Although greater incorporation of passive safety features, if undertaken with care and rigor, could be an advance in reactor design philosophy, we are concerned with the constant references by advanced reactor advocates to the supposed "inherent safety" of their designs.

Regardless of the validity of claims about immunity to the meltdown accident scenario, this terminology of "inherent safety" has more rhetorical merit than technical content. It is fundamentally misleading to describe as "inherently safe" a technology which necessarily contains and produces such large amounts of extremely hazardous material as does nuclear power. Although it may be possible to design a reactor which renders certain accident scenarios virtually impossible— or to make reactors that are considerably safer *relative* to existing reactors—that does not mean that the technology *per se* can be considered to have acquired safety as an inherent characteristic.

As stated in a 1990 study by the Union of Concerned Scientists (UCS), which considered several advanced reactor designs:

> As a general proposition, there is nothing *"inherently"* safe about a nuclear reactor. Regardless of the attention to design, construction, operation, and management of nuclear reactors, there is always something that could be done (or not done) to render the reactor dangerous. The degree to which this is true varies from design to design, but we believe that our general conclusion is correct.[214]

This conclusion is not limited to groups such as the Union of Concerned Scientists, which maintain a healthy skepticism about nuclear power. A study conducted by Oak Ridge National Laboratory also has reached similar conclusions:

> A nuclear reactor can never be completely inherently safe because it contains large quantities of radioactive materials to generate usable heat-energy; but nuclear reactors can be made

---

[214]UCS 1990, p. 3-5.

inherently safe against some types of events and have characteristics which limit consequences of certain postulated accidents.[215]

These cautionary statements raise another crucial concern: the possibility that in designing to eliminate certain now-commonly recognized accident possibilities, new accident scenarios will be unwittingly
introduced. As a survey of advanced designs by Britain's Atomic Energy Agency concluded:

> Safety arguments, in many cases, are very underdeveloped,
> making it difficult to gauge if the reactor is any safer than tra-
> ditional systems. [Advanced reactor] designers tend to concen-
> trate . . . on one particular aspect such as a [loss-of-coolant
> accident], and replace all the systems for dealing with that with
> passive ones. In so doing, they ignore other known transients
> or transients possibly novel to their design.[216]

This is an important warning. Nuclear technology is complex, and
it has taken many years of analysis and experience to even recognize
the existence or the possibility of some accident possibilities for the
four-decade-old light water reactor. The history of nuclear power development is replete with instances of incidents occurring at operating
power plants which had not previously been thought possible. This is
even true of the meltdown scenario discussed above, which was not
even recognized as a safety issue until the mid-1960s—over a decade
after the decision to build the Shippingport reactor. In view of this
history and the complexity of reactors, it would be prudent to antici-
pate that similar unexpected discoveries may be encountered in the
development of a new generation of reactors based on any new design.

The verification of the safety claims of any particular vendor, of
course, requires that the details of the design be made public so they
can be examined for potential safety flaws. Handwaving arguments
about general design features which are alleged to guarantee inherent
safety should not be allowed to substitute for actual design details and

---

[215]C.W. Forsberg et al., Oak Ridge National Laboratory, *Proposed and Existing
Passive and Inherent Safety-Related Structures, Systems, and Components (Building
Blocks) for Advanced Light-Water Reactors,* ORNL-6554, October 1989, pp. 1-2, as
cited in UCS 1990, p. 3-5.

[216]As reported in *Nucleonics Week* 1989, p. 9.

real-world data on actual components. To a large extent, however, the fine engineering details do not yet exist for designs that are not yet "construction ready."

The entire debate to date on the issue of the level of safety of new reactor designs has taken place largely on a theoretical level. While theoretical work is a necessary part of design, it cannot settle all essential safety questions by itself. Even the degree of relative safety of a reactor design is no easy matter to determine. Questions relating to the net level of improved safety are highly complex, and rely on substantial analysis of the fine details of design and experience accrued over time.

Safety uncertainties can never be fully resolved in advance, and will inevitably remain large until many years of operating experience have been acquired with advanced reactor designs. That is a crucial problem in the development of nuclear power. Operating experience is needed to make the right decisions about overall designs as well as critical detail, but getting that operating experience in itself involves non-negligible risks, at least if the scale of reactors is anywhere close to those required for large-scale commercial power generation. The only approach that could resolve this aspect of the problem of nuclear power is to study designs on paper thoroughly and then to acquire long experience with small scale devices, much in the manner that small-scale models of airplanes are tested extensively in wind tunnels prior to construction of full-scale prototypes.

Accidents and Nuclear Technology

Three major reactor design concepts have been put forward since the start of the nuclear era that have been implemented in commercial nuclear power:

- water-moderated and water-cooled reactors (light or heavy water);

- graphite-moderated (water-cooled or gas-cooled);

- unmoderated liquid sodium-cooled fast neutron reactors.

As with any technology, there have been a variety of problems in the development and implementation of nuclear power plants which have led to improved safety features. Some malfunctions were the result of experiments to test reactor designs, as was the case with the partial meltdown of the EBR I reactor in Idaho. Table 7 shows a list of some reactor accidents, including the major known ones.

Despite the considerable progress in understanding reactor safety over five decades (including experience with Manhattan Project reactors), the potential for catastrophic accidents continues to exist. A major reason is that nuclear power reactor designs were selected too quickly on the basis of energy, economic, military, and political criteria that did not give sufficient weight to the problems associated with catastrophic accident possibilities.

Light water reactors, by far the most common design today, were the simplest for the U.S. to build in the short term and hence gave the largest propaganda advantage to the United States during the Cold War. But this meant that the laboratory and theoretical work that was needed to understand the most severe accident, the loss of coolant from the reactor core, was completed over a decade after the 1954 decision to build Shippingport. By that time, the investment in the light water reactor was so great that the main reaction of the AEC was to try to cover up or downplay the seriousness of the problems.

There are at least three questions pertaining to catastrophic nuclear power plant accidents that are germane to the evaluation of the soundness of nuclear technology as a choice for future energy supply:

1. Is it possible to learn enough from non-catastrophic accidents in small-scale plants to prevent future catastrophic accidents in large-scale ones?

2. Is the scale of the accident such that the ill-effects could far exceed the benefits of any economies to be gained from nuclear energy versus some other energy choice?

3. Which generations would pay the price for the accident consequences—that which got the energy benefits or future generations?

Similar questions can also be asked about other technologies. We will address some aspects of this issue in the concluding chapter of this report. Let us first examine the Chernobyl accident, by far the worst in the history of nuclear power, for the lessons it might have to offer.

Table 7: Some Reactor Accidents

Reactor Type	Location	Accident Type	Year	Iodine-131 Release (curies)	Comments
Graphite-moderated, gas-cooled	Sellafield, Britain	graphite fire	1957	20,000	
Graphite-moderated, water cooled	Chernobyl, Ukraine	supercriticality, steam explosion and graphite fire	1986	7 million, perhaps far greater (see text)	Safety experiment went awry; total release 50 to 80 million curies or more; potential for continuing large releases exists
Sodium-cooled fast breeder	Lagoona Beach (near Detroit), U.S.	cooling system block, partial meltdown	1966	release confined to the secondary containment	Reactor was being tested for full power, but did not reach it; four minutes from indication of negative reactivity to meltdown
Sodium-cooled fast breeder	Monju, Japan	major secondary sodium leak	1995		Secondary sodium was not radioactive; reactor was in test phase; extensive sodium contamination in plant
Light water reactor, PWR type	Three Mile Island (near Harrisburg), U.S.	cooling system failure, partial meltdown	1979	13 to 17	Secondary containment prevented release of millions of curies of I-131; accident developed over several hours
Light water reactor, BWR type	near Idaho Falls, U.S.	accidental supercriticality followed by explosion and destruction of the reactor	1961	80	Small U.S. Army experimental reactor using HEU fuel; 3 operators were killed

Heavy water-cooled and -moderated	Chalk River, Canada	lack of coolant for a fuel element	1958	radioactivity contained within building	Highest worker dose 19 rem
Heavy water-moderated, light water-cooled, experimental	Chalk River, Canada	inadvertent supercriticality and partial meltdown	1952	"There was some release of radio-activity"	President Jimmy Carter helped in the clean-up
Heavy water-moderated and -cooled, CANDU type	Narora, Rajasthan, India	turbine fire; emergency core cooling system oper-ated to pre-vent system meltdown	1993	apparently no release of radioactivity	

Sources: Chernobyl: NRC 1987 and Medvedev 1990; Sellafield: Makhijani et al., eds. 1995, Chapter 8; Three Mile Island: TMI Commission 1979; Lagoona Beach (Fermi-I): Alexanderson, ed. 1979 and Fuller 1975; Idaho: Horan and Gammill 1963 and Brynes et al. 1961; Monju: press reports; Chalk River: John May 1989 and Weinberg 1994; Narora, press reports.

1. Chernobyl

On April 25, 1986, the operators of the Chernobyl Unit Number 4 were scheduled to perform an experiment designed to test an aspect of the safety of the RBMK design. The experiment was delayed for a number of reasons, including difficulty in stabilizing reactor power level. The operators decided to proceed with the test at 1:22 A.M. on the morning of April 26. Thirty seconds after the test began, an auto-matic computer printout indicated unsafe conditions, requiring the re-actor to be shut down immediately.

There followed a runaway supercriticality which greatly increased the power level, heated up the reactor, and increased the steam pressure in it to such high levels that it exploded, blew off the top of the reactor, and destroyed it. *Less than 90 seconds had elapsed between the com-puter warning to shut the reactor and the total destruction of the re-actor.*

Thirty fires were ignited in the reactor core and in other parts of the power plant, including the turbine building. Fire fighters arrived at the scene an hour-and-a-half later. They extinguished fires other than those in the reactor core relatively rapidly, but the reactor graphite fire lasted for ten days. Radioactivity releases went on for months after the fire had been extinguished.[217]

It was one of the two worst industrial disasters in human history, the other being the December 1984 disaster at the Union Carbide plant in Bhopal, India, during which deadly methyl isocyanate gas was released. In both accidents, hundreds of thousands of people were affected during the accident and in its aftermath. Thousands died on the night of the Bhopal catastrophe; in the case of Chernobyl, the immediate toll has officially been reported as 31, which is on the order of a hundred times lower. But the affected population increased dramatically in the aftermath of Chernobyl—130,000 people were evacuated, including the entire population of 45,000 in the town of Pripyat. More were evacuated subsequently, and hundreds of thousands of workers and soldiers were pressed into entombing the leaking reactor, digging up and burying vast quantities of highly contaminated soil, and performing other clean-up jobs.

Official estimates put the cumulative release of radioactivity between April 26 and May 6, when the fire was put out, at about 80 million curies. Of this total, 45 million curies are attributed to xenon-133, 7.3 million to iodine-131, 1 million to cesium-137, half-a-million to cesium-134, and 220,000 to strontium-90.[218]

These official Soviet estimates are misleading and understate the actual extent of the releases. For instance, the release estimates are adjusted for decay to ten days after the accident began. Xenon-133 has a half-life of 5.27 days and most of it was emitted early on in the fire. On this basis, the actual amount in the fallout cloud as it passed over communities was considerably greater. Similarly, iodine-131 has a

---

[217]Besides Medvedev 1990, Chapter 1, see also NRC 1987, Chapter 4, for a description of the details of the reactor accident and events leading up to it. For interviews with workers who were in the plant at the time of the accident and experienced some of its effects, see Chernousenko 1991, Chapter 2.

[218]NRC 1987, Table 6.3, p. 6-6. The radioactivity release figures in this table do not match up with those in Table 6.2, p. 6-4. The latter total, also supposedly decay-corrected, is 50 million curies, given by day of release, rather than broken down by radionuclide. Both tables give figures as calculated by Soviet authorities and presented by them to the International Atomic Energy Agency (IAEA).

half-life of 8.05 days and far more of it was deposited on grazing lands than indicated by the decay-corrected estimate of 7.3 million curies.

Zhores Medvedev, the Soviet scientist who first reported on the other nuclear catastrophe in the Soviet Union, the explosion in a high-level waste tank at Chelyabinsk-65 in 1957,[219] states in his study of the Chernobyl accident that the official figures for radioactivity releases include only the amounts deposited inside the former Soviet Union and do not take into account the much larger deposition of some radionuclides, such as iodine-131 and cesium isotopes outside Soviet territory. According to his analysis, this is because the Soviet government did not want to acknowledge "any liability for radioactive contamination of the environment in other countries" and hence it insisted that "the amount [deposited outside the Soviet Union] was negligible."[220] Medvedev estimates that releases of radioiodine and radiocesium were about three times higher than the official estimates cited above.[221]

One of the most important, unanticipated features of the Chernobyl accident was the ten-day duration of the fire, accompanied by a correspondingly long time during which large releases of radioactivity continued. As Medvedev points out, the modeling of nuclear power plant accidents generally assumes a single, short-term release of radioactivity. Weather conditions during such short releases can reasonably be assumed to be constant. As a result, severe accidents are assumed to have a fallout trace that forms a single elongated, cigar-shaped pattern, much like the typical fallout pattern from a nuclear bomb explosion near ground level. This assumption is sometimes valid. It was, for instance, the pattern of radioactivity released as a result of the 1957 Soviet explosion in a tank containing highly radioactive waste. But it was not valid for the Chernobyl accident.

During the ten days of the fire, which was accompanied by huge releases of radioactivity, wind directions and the weather changed many times. As a result, large, widely scattered areas in many compass directions were affected. Rainfall in some areas during this prolonged period created hot spots of radioactivity in three states of the Soviet

---

[219]IPPNW and IEER 1992, Chapter 4.

[220]Medvedev 1990, pp. 77-78. The rest of the section on Chernobyl is based on this source, unless otherwise mentioned.

[221]Medvedev 1990, p. 78.

Union, now separate countries: Ukraine, Belarus, and Russia. Countries far beyond the Soviet Union were also affected. Europe was especially affected by the fallout, and levels of iodine-131 in milk exceeded officially permissible levels in many countries. Every country in the northern hemisphere received some fallout from the accident.

An "exclusion zone" 30 kilometers in radius was established and, after delays, 130,000 people were evacuated. Agriculture and commercial activities were also prohibited in the area. But the actual area that was contaminated and the number of people affected was far larger. There were hot spots as much as 100 to 300 kilometers from the accident that had radiation levels on the order of one thousand times above natural background. Long-lived biologically sensitive radionuclides, notably cesium-137 and strontium-90, were deposited in large quantities. Figure 8 shows a map of contamination around Chernobyl.

Iodine-131 concentrates in milk. When this milk is consumed, it concentrates in the thyroid glands, especially affecting children. After the iodine-131 decayed away in a few months (ten to 20 half-lives), milk produced in the contaminated regions continued to be affected by cesium-134 and cesium-137 contamination. The ill-effects of cesium-137 will last for a hundred years or more.[222] There was a ban on open market milk sales in several regions, affecting 20 to 25 million people for more than a year after the accident.[223] Even with these extensive measures, milk production was not halted in all contaminated regions. Some people in the most rural areas immediately around the plant consumed contaminated milk in the aftermath of the accident at a time when sales of such milk had been banned in Kiev. Cesium-137 contamination of milk will continue for many decades.

---

[222] Cesium-134 and cesium-137 gave half-lives of about two years and 30 years, respectively.

[223] Medvedev 1990, p. 111.

Figure 8: Contaminated Areas Around Chernobyl

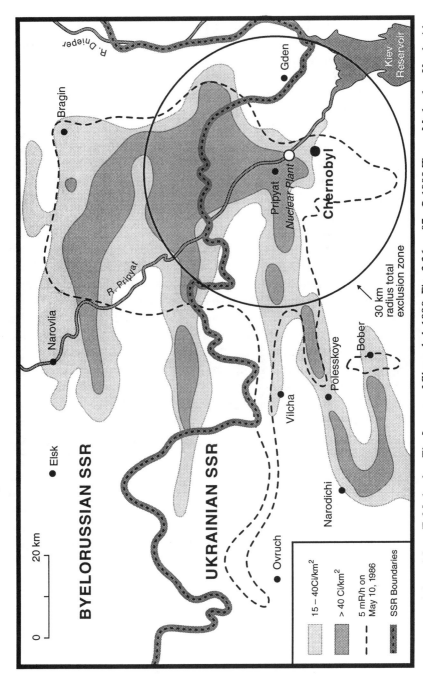

Source: Adapted from Z. Medvedev, *The Legacy of Chernobyl*, 1990, Fig. 3.36, p. 87. © 1990 Zhores Medvedev. Used with permission of W. W. Norton and Blackwell Publishers. *Note:* In legend, Ci = curies; mR/h = milliroentgens per hour.

The region around Chernobyl consists largely of swamps and soggy forest land. Much of the land has been reclaimed for agricultural use, the dominant use at the time of the accident being cattle grazing. The prevalent ecological conditions are conducive to retention of cesium and to its rapid transfer to plants. As a result, agriculture was affected over a vast region. The most immediately affected area was the 30-kilometer radius exclusion zone in which 70,000 hectares (175,000 acres) of fields, grazing land, and vegetable and fruit gardens were abandoned. In June, there was a further evacuation of 113 villages outside the exclusion zone in Belarus and Ukraine. Between 100,000 and 150,000 hectares (250,000 to 375,000 acres) of agricultural land were abandoned.

Levels of cesium-134 and cesium-137 contamination are especially important as criteria for suitability for agricultural use. Medvedev estimates that "if international standards were being applied for the use of agricultural land, nearly one million hectares would be considered lost for a century, and about two million hectares would be lost for 10-20 years."[224] There have been anecdotal reports of large increases in farm animals born with genetic defects. At one collective farm, 27 abnormal calves were born in the year after the accident while none had been reported in the five years preceding it. The number of suckling pigs with genetic defects increased from three cases in five years to 64 in one year.[225]

Most contaminated agricultural land continues to be used for farming. Indeed, many people who were evacuated from severely contaminated areas have returned to them due to economic problems in the areas to which they were relocated and the wish of many older people to live and die at home.

A large amount of agricultural produce in Europe had to be dumped due to contamination from fallout. For instance, most vegetables in the region around Munich were destroyed because they had become contaminated with iodine-131. The southern portion of the former West Germany was more contaminated than the rest of it. There were also severe restrictions on agricultural activities, including sales of meat from three million sheep and lambs in northwestern England and the neighboring portions of Scotland and northern Wales, which were af-

---

[224]Medvedev 1990, p. 110.
[225]Medvedev 1990, p. 116.

fected by rain-out of radioactivity when the fallout cloud passed over them.

Health Effects

Several categories of people have been and will be affected adversely by radiation doses from the Chernobyl accident:

- *Workers in several categories*: those who were in the plant, put out the fire, cleaned up afterwards, built the concrete structure around the burned-out exploded reactor, and monitored or otherwise performed supporting functions in contaminated areas.

- *The people in the region whose land and homes became contaminated.*

- *People in the regions who consumed, continue to consume, or will consume contaminated food and/or water.*

- *People who received radiation doses from the fallout*, with the highest doses generally being in the former Soviet Union and Europe.

The assessments of adverse health effects from the accident have varied widely. Official reports have tended to concentrate on the 31 workers who died of severe radiation exposure. But this position ignores the far greater numbers of people who were exposed to considerable levels of radiation, became ill in the months and years that followed the accident, and who have an elevated risk of various radiogenic diseases in the years to come. It also does not take into account the effects of the accident for decades to come.

One complicating factor in assessing the health risks due to the accident has been the severe deterioration, bordering in many areas on collapse of social services, including health delivery services in the former Soviet Union. As a result, the increases in diseases and death due to radiation exposure are mixed up with those arising from the general deterioration in medical care and economic conditions.

Some indication of the potential health damage can be obtained by looking at the radiation doses. The range of exposures of the people who lived in the exclusion zone was generally of the same order of magnitude as the survivors of Hiroshima and Nagasaki—that is, about one rem to several tens of rems external gamma radiation. In addition, people were exposed to beta radiation and internal doses from various radionuclides, such as iodine-131 and cesium-137. The officially estimated cumulative population dose for the 135,000 people who were initially evacuated (with delays) is estimated at 1.6 million person-rem. Applying a risk factor of 0.0004 cancers per person-rem to this dose yields an estimate of 640 fatal cancers.[226]

Medvedev has pointed out that the official dose estimate includes only external radiation. It does not include doses from consuming contaminated food, such as milk continuing cesium isotopes and iodine-131. It is now clear that internal exposures are a significant factor in long-term effects of the accident. Thyroid diseases, including thyroid cancer in children generally attributed to the consumption of milk contaminated with iodine-131, have registered huge increases in the fall-out areas. Ten to one hundred-fold increases in thyroid cancer among children in the affected region have been reported.[227] Over the decades tens of millions of people will have been put at significantly increased risk, and it is reasonable to assume that many will die as a result. The poor state of both medical monitoring as well as curative medicine in the former Soviet Union means that medical systems are not likely to record many of these deaths as having been related to the Chernobyl accident. But that cannot negate the documented magnitude of the immense contamination and risk to which the present and future generations living in tens of thousands of square kilometers of highly contaminated land are being, and will continue to be, exposed.

The number of deaths from increased exposures even in the far off contaminated regions in the European Community (EC) are projected to be large. The British National Radiation Protection Board estimates that up to 1,000 additional cancer deaths will occur in the EC region due to radiation doses from radiocesium and iodine-131. Medvedev considers this a "minimal assessment."[228]

Medvedev has cited the entire range of estimates for cancer death estimates that have been made. The lowest estimates are 200 to 600 additional cancer deaths in the former Soviet Union, while the highest

---

[226]For a discussion of cancer risk factors per unit of radiation, see Makhijani et al., eds. 1995, pp. 19-21 and 72-74.

[227]Baverstock et al. 1992; and Likhtarev et al. 1995.

[228]Medvedev 1990, p. 165.

estimate is 280,000 additional cancer fatalities worldwide.

These estimates do not include adverse health effects on workers and soldiers who were the clean-up crews and hence among the most severely affected. There are no systematic records of their exposure or even of how many of them were involved. Medvedev quotes an eyewitness account of the working conditions of the soldiers who did the clean-up work in the immediate aftermath of the accident:

> I saw soldiers and officers picking up graphite [ejected from the reactor core by the explosion] with their hands . . . There was graphite lying around everywhere, even behind the fence next to our car. I opened the door and pushed the radiometer almost onto a graphite block. Two thousands of roentgens an hour . . . Having filled their buckets, the soldiers seemed to walk very slowly to the metal containers where they poured out the contents, You poor dears, I thought, what an awful harvest you are gathering. . . .
>
> The faces of the soldiers and officers were dark brown: nuclear tan.[229]

Medvedev estimates that the radiation tan on the soldiers' faces indicates skin doses of 400 to 500 rem, that many of them suffered from acute exposures, and that some died as a result. No records have been kept or made public, at any rate, of the numbers of soldiers involved in such activities or of their exposures.

Large numbers of workers were also exposed to high levels of radiation in the years that followed when a concrete "sarcophagus" was built around the burned-out reactor building to try to encase the radioactivity. Two hundred thousand men, working very short shifts, were involved in its construction. The radiation levels were extremely dangerous, with the most radioactive areas measuring between 5,000 and 20,000 rads per hour. The sarcophagus was built in the hope that it would contain the radioactivity for an extended period. But it has already deteriorated considerably and new measures to contain the radioactivity appear to be necessary. There is no consensus on the appropriate approach to contain the enormous amount of radioactivity in and under the building, but whatever measures are taken, they will be costly. If measures are not taken, the costs, in terms of contamina-

---

[229] Grigory Medvedev, quoted in (Zhores) Medvedev 1990, pp. 167-168.

Chernobyl Sarcophagus. The April 1986 reactor explosion and fire destroyed the reactor which was then entombed in this sarcophagus. The sarcophagus has deteriorated considerably. (Credit: James Lerager)

tion of important sources of water supply of the region, could be far higher.

The overall costs of the Chernobyl accident are so vast and extend over so many generations that they are impossible to calculate. The official calculation of eight to 11 billion rubles (1988), or roughly ten to 15 billion dollars. But any evaluation is complicated by the fact that a large number of clean-up workers are neither being followed or treated. It is also very difficult to quantify the economic and social losses caused by the uprooting of hundreds of thousands of people. Further, the high radiation doses received by many mean that problems other than cancer are also likely to occur. For instance, diseases induced by the weakening of immune systems of clean-up workers and off-site populations who received high radiation doses could cause large health and economic impacts. But, given the state of the health delivery systems, they would be difficult or impossible to detect. Finally, the negative impact of Chernobyl on the electricity systems of the former Soviet Union is still being felt and enormous costs loom in terms of preventing the spread of radioactivity from the reactor, preventing accidents at other reactors of the same design, and replacing reactors generally considered to be unsafe well before their design lifetimes. The costs of replacing electric-generating capacity not provided by RBMK reactors, which are generally considered to be far too dangerous in the West, could by itself run into tens of billions of dollars.

2. Some Lessons of the Chernobyl Disaster

The most important and tragic lesson of the Chernobyl accident is the most severe kind of nuclear power accident can actually happen. Nuclear power technology is unforgiving. It has often been stated by proponents and opponents alike that it does not allow room for mistakes. Design, management, and operator errors have typically combined to yield accidents; in many cases, these same features have also helped limit the damage. In the case of Chernobyl, the factors propelling the situation toward a major accident completely overwhelmed any checks in the system.

It is generally agreed that accidents on the scale of Chernobyl or worse are more probable in the former Soviet Union and Eastern Europe, but they are also possible elsewhere. That potential has been demonstrated events such as the 1979 Three Mile Island accident and the British Windscale reactor fire in 1957. The scale and the irremediable nature of the damage from Chernobyl leads to a crucial question: Is it possible to design nuclear reactors that would not be subject to accidents of such catastrophic magnitude? This is not the same as ruling out all accidents, which is clearly impossible with any technology. It is merely to ask whether the damage can be limited so that it is at least remediable in its worst aspects.

As we have discussed, current nuclear power plant designs do not meet this goal. LWRs, graphite-moderated reactors, or sodium-cooled reactors in the West all have vulnerabilities in design and/or operation that could lead to severe accidents. The record shows that the probabilities of catastrophic accidents are lower in the West than in the former Soviet Union. But this is an inadequate response, given the nature of the consequences and the fact that energy alternatives that would avoid catastrophic accident potential are available.

We can grant that the safety of nuclear power plants in the United States has improved over the decades, as public vigilance and the Three Mile Island accident have forced the manufacturers to conform to stricter safety standards. However, these efforts cannot negate the fact that current power reactor designs are vulnerable to catastrophic accidents. Chernobyl demonstrates that the effects of such accidents are as devastating as they are irremediable. In this context, it is well to recall a criticism of nuclear power plant safety efforts made by Nobel laureate physicist Hannes Alfven in 1972:

> The reactor constructors claim that they have devoted more effort to safety problems than any other technologists have. This is true. From the beginning they have paid much attention to safety and they have been remarkably clever in devising safety precautions. This is . . . not relevant. If a problem is too difficult to solve, one cannot claim that it is solved by pointing to all the efforts made to solve it.[230]

---

[230]Alfven 1972.

CHAPTER 8:
PLUTONIUM DISPOSITION, MILITARY TRITIUM, AND COMMERCIAL REACTORS

In the early years of the Cold War, many nuclear power proponents proposed that military plutonium production be used to subsidize commercial nuclear power plants. After the end of it, there are proposals that the use of excess military plutonium in reactors would be a suitable route for the development of new commercial reactor designs or for providing subsidized fuel to existing power plants.

There is also a possible new production connection between the military and commercial nuclear industry. Tritium, a radioactive isotope of hydrogen, is now used in most or all nuclear weapons to increase the efficiency with which plutonium and highly enriched uranium are used during the nuclear explosion. In the United States, tritium was produced in some of the same military reactors that were used to make military plutonium.[231] Now that the military reactors are all shut due to safety concerns, the potential exists for a new military-commercial nuclear connection in the form of tritium production in existing or new power-producing reactors.

The issue of disposition of plutonium has been extensively dealt with in the public literature, including major reports in 1994 and 1995 by the National Academy of Sciences and a book published in 1995 by the Institute for Energy and Environmental Research.[232] The problem of tritium production is mainly related to U.S. nuclear strategic posture. We will summarize the issues here, mainly focusing on the proliferation and environmental issues that are raised by recreating a military-civilian connection in the U.S. nuclear industry.

---

[231]A small amount of this tritium was also sold for commercial applications. Canada is a major source of tritium used in civilian industry.

[232]NAS 1994; NAS 1995; and Makhijani and Makhijani 1995.

Tritium production and plutonium disposition have been joined in some proposals, such as those for a new "triple play" reactor which would "burn" MOX fuel, generate electricity, and produce tritium.

Plutonium Disposition[233]

The retirement of warheads in the United States and Russia at the end of the Cold War has generated surpluses of military plutonium and highly enriched uranium. The secure storage of these materials has become a grave concern due to the collapse of the Soviet Union. In principle, highly enriched uranium could be mixed with natural uranium, depleted uranium, or slightly enriched uranium and hence be converted into low enriched uranium. This could be stored or used in existing light water reactors as a substitute for fuel derived from uranium mining, processing, and enrichment.[234]

But the disposition of surplus plutonium is a far more complex problem. At the dawn of the nuclear era, as we have discussed, plutonium was envisaged as the fuel that would be at the center of an age of nuclear energy. But uranium turned out to be far more plentiful and far cheaper than predicted. At the same time, the costs, environmental dangers, and safety issues associated with breeder reactors and reprocessing turned out to be high, relative to the use of uranium alone without these technologies. Plutonium, although an energy resource in principle, had become an economic liability. Potential proliferation concerns had also made it into a security liability.

The high cost of processing and fabricating plutonium into fuel makes it more expensive to use as a fuel in commercial reactors, whether by itself (in breeder reactors) or as a partial substitute for uranium (in the form of mixed oxide fuel, known as MOX fuel). The majority of independent studies in the U.S. done in recent years have concluded that plutonium is not an economical energy resource. But the need to put surplus military plutonium into a non-weapons-usable form is urgent, because this would reduce the risk of black market sales

---

[233]This section is based on Makhijani and Makhijani 1995, unless otherwise noted.

[234]We will not discuss the problems involved with disposition of highly enriched uranium in the context of this study of reactor development, since the two issues are relatively independent. For a discussion of disposition of highly enriched uranium from the U.S. nuclear weapons program, see Makhijani and Makhijani 1995, Chapter 7.

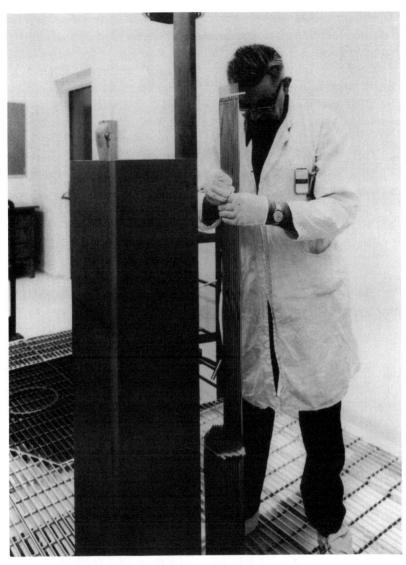

Mixed oxide nuclear fuel assembly. Fuel pellets with a mixture of plutonium and uranium oxide are lowered into a hexagonal array for the liquid
metal fast breeder reactor under construction in the 1970s at the Hanford
Engineering Development Laboratory at Richland, Washington. (Credit:
Battelle Northwest Photo)

of plutonium, notably that of Russian origin. The process should also make the reuse of plutonium in nuclear weapons difficult and costly. Many operators of existing commercial reactors as well as new reactor vendors have expressed an interest in getting government subsidies to "burn" (that is, fission) plutonium as a nuclear fuel in reactors. This would put plutonium in the form of spent fuel, which cannot be used in nuclear weapons without reprocessing. However, a substantial fraction of the plutonium would still remain after one pass through a reactor. Every technology to completely burn plutonium requires some kind of reprocessing technology, with attendant proliferation, environmental, and cost concerns.

Finally, a "triple play" reactor has also been proposed by some in the nuclear industry. Such a reactor would generate electricity, produce tritium, and also burn some of the excess plutonium from the nuclear weapons program. This option is not now officially favored by DOE, but it still has its proponents in the U.S. Congress and in industry.

The principal criteria for disposition of plutonium as discussed in Makhijani and Makhijani 1995 are:[235]

> 1. *Security aspects*: The treatment, storage, and disposal of plutonium as a waste must be such that the difficulty of plutonium re-extraction from the waste is as close to new plutonium production and separation as possible.
>
> 2. *Time frame:* Putting plutonium into non-weapons-usable form as soon as possible (compatible with protection of the environment and of worker and community health) is crucial in light of the situation in the former Soviet Union. Russia is unlikely to act without the U.S. doing so also.
>
> 3. *Accident risks*: The risk of catastrophic accidents, resulting in the dispersal of plutonium or accidental nuclear or non-nuclear explosions, must be evaluated for each option.
>
> 4. *Health, environmental protection, and safety*: The option chosen should be compatible with compliance with all applicable environmental, health, and safety laws and regulations. It should take account of the reality that increased handling, processing, and transportation entail additional new environmental risks, and that some of these new risks may offset existing risks from storage.

---

[235]Makhijani and Makhijani 1995, pp. 19-20.

5. *Potential for encouraging plutonium production:* Some disposition options involve the use of reprocessing technologies
and/or of facilities to fabricate fuel containing plutonium. Hence
there is a need to consider the potential for a U.S. choice of a
disposition option to entrench the separation and use of plutonium in other countries.

6. *Cost:* It is important to compare the costs of various disposal
options for plutonium, Although in light of the immense security
risks involved, this is a secondary issue.

Compatibility with final disposal is also a concern. For instance,
issues such as durability of the waste form in the repository setting and
inadvertent criticalities in the repository need to be addressed.

A large variety of options have been put forward to put plutonium
into non-weapons-usable forms. They can be classified into three basic
approaches:

- Disposing of plutonium without processing, for example, by disposal in deep boreholes, by sub-seabed disposal, or by shooting
  it into space;

- Using plutonium as a fuel in a reactor;

- Immobilizing plutonium mixed with contaminants, such as fission products in glass logs or other waste forms so that it is
  difficult to use it in weapons or to steal it.

Many of the reactor options are not regarded with favor because
they do not adequately meet plutonium disposition criteria set forth
above. Table 8 on the next page shows these rejected reactor options.

Two studies by the National Academy of Sciences on plutonium
disposition have recommended that excess military plutonium be put
into non-weapons-usable form in one of the following ways:[236]

- Burned in existing reactors as a mixed oxide fuel;

- Vitrified with fission products so that the radioactive glass logs

---

[236]NAS 1994 and NAS 1995. The NAS also recommended further research on and
investigation of deep borehole disposal as a potential option. The boreholes would be
much deeper than the repository now contemplated for high-level radioactive waste.

Table 8: Rejected Minimized Accessibility Plutonium
Disposition Options

Disposition Option	Principal Reasons for Rejection
New burner reactors, no reprocessing	Long time frame; licensing uncertainties.
New thermal reactors with reprocessing	Encourages reprocessing and hence undermines non-proliferation goals; long time frame.
Advanced Liquid Metal Reactor (ALMR)	ALMR can be used to breed plutonium; most proposals for its use also require a new reprocessing technology (pyroprocessing); long time frame; undermines non-proliferation goals.
Pyroprocessing without ALMR	Promotes development of a new reprocessing technology under the guise of plutonium disposition; undermines non-proliferation goals.
Nuclear explosion in an underground cavity	Extensive and unacceptable environmental damage; undermines the non-proliferation goal of stopping nuclear explosions.
Subcritical reactor with proton accelerator	Involves development of a reprocessing technology and hence undermines non-proliferation goals; long time frame; high technical uncertainty.

Source: Makhijani and Makhijani 1995

approximate spent fuel in terms of the difficulty they would pose for theft or for re-extraction for use in weapons.

The first option has many variants, but most of them would reconnect the nuclear power industry with U.S. nuclear weapons establishment.[237] The most important proliferation problem associated with use of plutonium in reactors is that it would create an infrastructure for long-term use of plutonium in the civilian economy. Once the capital investment has been made at government expense in costly facilities, notably a plant for fabricating mixed uranium oxide-plutonium oxide fuel (MOX fuel), civilian industry will be tempted to go on demanding

---

[237] A government-owned reactor for plutonium disposition has also been proposed, with the express idea of avoiding connecting military issues with commercial nuclear power. See NAS 1994; and Makhijani and Makhijani 1995.

subsidized government-supplied fuel. The use of MOX fuel in the United States would also make it very difficult for the U.S. government to persuade other countries, such as Russia, Japan, and India, to stop reprocessing and accumulating separated plutonium (see the next chapter). Finally, creating a government-financed infrastructure for using plutonium as a fuel could also encourage companies or government agencies to seek money for breeder reactors, even though these reactors remain highly uneconomical, continue to be plagued with technical and safety problems, and pose special proliferation risks. Burning plutonium in foreign reactors—for instance, in Canada or Germany—has also been proposed, each country possessing diverse advantages extolled by their proponents. However, they all share the proliferation disadvantage of creating and using a MOX fuel infrastructure.

The pressure to build one or more new reactors continues. As noted above, there are those who favor a new "triple" play reactor, whether this be of the light water variety or an MHTGR. The latter reactor is also being promoted for sale in Russia as a method for plutonium disposition there. Due to the severe financial difficulties in Russia, the proposal appears to be a non-starter unless the U.S. government provides a substantial portion of the funds to build the reactor.

While there is no really good option for converting plutonium into non-weapons-usable form, vitrification is by far the most preferable option when such considerations as speed, environmental protection, and non-proliferation are taken into account. Table 9 on page 172 shows some approaches to vitrification, along with their advantages and disadvantages.

There are numerous variants of these options. For instance, an initial vitrification of plutonium alone would rapidly put it into non-weapons-usable form. This could then be followed by vitrification with fission products.

Military Tritium[238]

In the 1980s, the DOE planned to upgrade some of its aging reactors that had been shut down due to safety concerns and to build new military production reactors. The primary purpose was to produce tritium for use in nuclear weapons. Tritium, an isotope of hydrogen, is a

---

[238]This section is based on Zerriffi 1996, unless otherwise specified.

Table 9: Comparison of Vitrification Options

Option	Advantages	Disadvantages
1. Vitrification of plutonium alone	Simplest and most rapid option	Least technical difficulty for plutonium re-extraction; low resistance to theft
2. Vitrification of plutonium with fission products	Highest initial proliferation-resistance in regard to both difficulty of theft and of re-extraction	May hamper global agreement on an interim halt to reprocessing; likely to take the longest time; in a few centuries, proliferation resistance declines to approximately that of vitrification of plutonium alone.
3. Vitrification of plutonium with actinides or rare earths	Moderate-to-high technical proliferation resistance; can be done rapidly; durable proliferation resistance	Low resistance to theft; re-extraction less difficult than with Option 2.
4. Option 3 with a gamma-emitting canister	High technical proliferation resistance; can be done rapidly; durable proliferation resistance; high resistance to theft	Re-extraction less difficult than with Option 2.

Source: Makhijani and Makhijani 1995, p. 46.

radioactive gas that decays with a half-life of 12.3 years. Modern nuclear warheads contain tritium in order to increase their explosive power (called yield) and the efficiency of use of the plutonium or highly enriched uranium in the primary stage of the warhead. Tritium can be produced in reactors or accelerators by bombarding lithium-6 or helium-3 (both non-radioactive elements) with neutrons.

In the early 1990s, nuclear power plant vendors saw military tritium requirements as a prime opportunity for getting military orders to make up at least partly for the moribund commercial side of the business. Reactor vendors, notably General Atomics with its MHTGR, also thought it was a good opportunity to market new reactor designs and get the government to pick up at least some of the development costs, the way it had the first time around in the 1950s. These desires received a boost when the plans to upgrade DOE's reactors at the Savannah River Site in South Carolina ran into such high costs and continuing safety and environmental issues that they were canceled by 1993.

By that time, however, the Cold War had ended and the Soviet Union had disintegrated. Agreements to reduce strategic nuclear warheads were being implemented. In the wake of the attempted military coup in the Soviet Union in 1991, President Bush ordered the withdrawal of most U.S. tactical nuclear warheads from deployment in the hope that the Soviet Union would follow suit, which it did. He was concerned about potential black markets in nuclear weapons or weapons-usable materials—which has since come to be known as the "loose nukes" problem. But the Pentagon has decided to keep about 5,000 warheads in the U.S. arsenal, which would require tritium production to resume by the year 2011, according to DOE calculations. Smaller numbers of warheads, desirable for a host of reasons, including fulfilling commitments under the Nuclear Non-Proliferation Treaty, could extend this date for a considerable period. Table 10 shows the dates for the resumption of tritium production as a function of the arsenal size and (for low arsenal sizes) of commercial tritium sales.

The DOE has decided to investigate an accelerator and the use of existing commercial reactors for tritium production. As noted above, however, there are still proponents of a new reactor that would serve as a "triple play" reactor, one of whose functions would be military tritium production.

In 1997, the Department of Energy in collaboration with the federally owned utility, Tennessee Valley Authority (TVA), began to irradiate a test target assembly at the Watts Bar reactor. The tests are supposed to be completed in 1999. The TVA has also proposed that its two incomplete nuclear reactors near Scottsboro, Alabama, be completed with government funds at an estimated cost of $2 billion and used for tritium production. This has aroused opposition on non-proliferation grounds, since it would breach the wall that has traditionally separated commercial and military nuclear reactors in the United States (unlike several other countries).

Concluding Observations on the Civilian-Military Nuclear Power Connection

Cost subsidization has come a perverse full circle from the late 1940s, when it was argued that military production of plutonium could

Table 10: Stockpile Levels (1,000 and above) and Tritium Requirements

Number of Warheads	Tritium Inventory Required (Kilograms in Warheads and Pipeline)	Year of New Tritium Production (See Note 1)	Comments
4700	28.5	2011	Approximately the size of the START II stockpile plus tactical and spare warheads.
3500	22	2015	Number of START II strategic warheads.
1000	12.1	2024	Current Tritium Pipeline
1000	7.0	2032	Tritium Pipeline reduced to 2.0 kilograms
500	3.5	2040	Could be stretched to 2044 without commercial sales
200	1.4	2047	2060 without commercial sales (See Note 2)
100	0.7	2050	2073 without commercial sales (See Note 2)
1	0.007	2054	2155 without commercial sales (See Note 2)

Source (including Note 1): Zerriffi 1996.

Notes: 1. The year in which new tritium production would be required is calculated using a decay rate of 5.5 percent per year, a 1992 tritium inventory of 88.2 kilograms, yearly commercial sales of 0.15 kilograms and five grams physically in each warhead for levels below 1,000 warheads. 2. The dates for how long the existing military tritium stockpile could be stretched without commercial sales are calculated only for stockpile levels below 1,000 because commercial sales at present levels would not significantly affect the date for stockpiles of 1,000 warheads or more.

subsidize the generation of civilian electricity. Now, in order to get government money, a considerable portion of the reactor industry istrying to convince government that it should get public funds for burning of excess military plutonium. There is a considerable conflict of interest involved since some of the corporate proponents of disposition of plutonium by using it as MOX fuel are the very same corporations that have a stake in the commercial nuclear industry. Prime among these corporations is Westinghouse, which runs Savannah River Site and the Naval Reactors Facility at the Idaho National Engineering Laboratory; it also manufactures commercial PWRs. These corporations also have the support of personnel in the field offices of the DOE, where professionals who have built their careers around plutonium production and processing are highly reluctant to acknowledge that plutonium is an economic liability rather than an energy resource.

The proposal by the TVA and DOE to produce military tritium in commercial reactors would constitute a major subsidy to the U.S. nuclear power industry at a time when reactors are being shut rather than built. The context of both the MOX subsidies and the tritium subsidies is the ongoing deregulation of the electric utility industry in the United States. Deregulation is exposing the true high costs of many nuclear reactors. These high-cost reactors are increasingly being seen as "stranded costs" since there would be no way to make a profit on such investments in competitive electricity markets.

There is also a renewed threat that reprocessing of civilian spent fuel will resume in the United States due to the serious problems in the repository program for spent fuel (see next chapter). There has been at least one specific proposal to begin such reprocessing at the Savannah River Site (see Chapter 6).

CHAPTER 9:
NUCLEAR POWER AND ENERGY POLICY

The history of the connection of commercial nuclear power to the military indicates that the link has been a damaging and costly one. But the overall question of the role of nuclear power in the United States, or in any other country, cannot be decided on the basis of the negative aspects of the military connection alone. A sound policy on nuclear power must be decided by considering the issues specific to commercial nuclear power within the context of an overall energy policy. Such a policy must necessarily take into account a host of technological, economic, environmental, political, and security factors. Since nuclear power is primarily oriented toward electricity generation, the role and evolution of electricity in energy supply and use also needs to evaluated.

We will consider the role of nuclear power within the United States, since that is the focus of this book. But we also consider briefly the global energy picture since this is important to energy and security issues in the United States. Further, the Article IV of the Nuclear Non-Proliferation Treaty (NPT) requires signatory nuclear weapons powers to share commercial nuclear technology with other signatory states, including non-nuclear signatories.

Table 11 below shows energy and electricity consumption in the United States, and the role of nuclear energy in it in 1992.

Just over one-third of U.S. energy consumption is used for electricity generation. The rest of energy supply goes into direct fuel use in industry, transport, and applications, such as space heating of homes and commercial buildings. The proportion of energy that goes into electricity increased from about 19 percent in 1960 to 26 percent in 1980 to over 35 percent in 1992. In fact, the fuel used by electricity sector accounts for the entire growth in energy use since 1973. Of course, this electricity is used, in turn, in homes, offices, etc. The net result is that the U.S. economy, like that of other countries, is becoming

Table 11: Energy Consumption and Electricity Generation
in the United States, 1992

Fuel	Amount Consumed (Exajoules)	Percent of Energy Use	Electricity Generation (Billion kWhe)	Percent of Electricity Generated from Fuel
Coal	19.9	22.9	1,576	56.3
Oil	35.3	40.6	89	3.2
Natural Gas	21.4	24.6	264	9.4
Nuclear	7.1	8.2	619	22.1
Hydroelectricity	3.0	3.5	239	8.5
Other	0.2	0.2	10	0.4
Total	**86.9**	**100**	2,797	**99.9**

Source: Statistical Abstract 1994.

Notes: 1. Percent totals may not add due to rounding.
2. Energy values in exajoules can be converted into British units of quadrillion Btu by dividing the exajoule figures by 1.055. One metric ton of U.S. coal on the average is approximately 25 billion joules. Therefore, one exajoule (which is equal to one billion billion joules) is equivalent to about 40 million metric tons of U.S. coal.

more electricity intensive. This trend would intensify greatly if road transport becomes largely electrified through the use of battery-powered (rather than methane- or hydrogen-powered fuel cells) vehicles.

The share of nuclear-generated electricity in the United States grew from under 2 percent in 1970 to over 22 percent in 1992. However, the number of nuclear plants operating in the United States has begun to edge downward from a peak of 111 in 1990-1991 to 109 in 1994. The number of operating reactors increased by one in November 1995, when the Watts Bar 1 reactor of the Tennessee Valley Authority went on line. Five other reactors have an official "deferred" status.[239]

While coal has retained its pre-eminent place as the source of fuel for electricity generation throughout the nuclear era, it should be kept in mind that 20 percent of the U.S. electricity supply contributed by nuclear power plants is a large amount of electricity both in absolute terms and in terms of the share of electricity supply. In some regions,

---

[239]NRC 1995, pp. 24, 30-40; and NEI 1996, p. 5.

such as Illinois, the share of nuclear-generated electricity is even larger. In many power supply systems, nuclear electricity constitutes such a large fraction of the generation capacity that it is essential to the overall stability of the electricity supply system.[240] For this reason alone, nuclear power plants cannot all be turned off overnight or even within a few years, however desirable it may be to reduce the generation of further weapons-usable materials and the risk of catastrophic accidents like Chernobyl. But it is possible to phase them out over a longer period.

There is an analogy between the build-up of greenhouse gases and reducing the emissions of ozone-depleting compounds. Human beings are affecting the composition of the atmosphere in ways that risk terrible and sustained harm to future generations. It is highly desirable to reduce carbon dioxide emissions by about 50 percent as soon as possible, so that anthropogenic emissions may correspond approximately to the apparent capacity of nature to absorb them. But the burning of fossil fuels cannot be cut by 50 percent within a few years because there is no social consensus to do so and because it would cause terrible adverse economic, social, and health consequences if attempted so suddenly. Even in the matter of protecting the ozone layer, which is essential to life on Earth, it took a huge, global diplomatic effort to phase out the production of the most harmful ozone-depleting compounds (called chlorofluorocarbons, or CFCs) in approximately one decade. But this occurred only after there was substantial agreement in the scientific community and among many governments, including the United States, that it was essential in principle to phase out CFCs.

World commercial energy consumption in 1992 was about 350 exajoules, with an additional 40 exajoules being consumed in the form of traditional energy sources, such as fuelwood. World electricity production in 1992 was about 12,000 billion kWhe. In 1993, nuclear power generation was 2,080 billion kWhe from 424 nuclear reactors in 33

---

[240]Nuclear power plants are "baseload" plants—that is, they are designed to contribute to that portion of the utility's power that is relatively constant throughout the day and over the seasons. This is because their capital cost is high relative to their fuel cost. "Peaking" capacity provides the daily and seasonal fluctuations. This is primarily oil and gas (and to some extent hydro), where fuel costs are high relative to capital costs. Baseload plants are particularly important to system stability, since they typically are large units, each of which contributes considerably more to overall system capacity than the smaller peaking units.

countries. Most nuclear generation, however, is in a handful of countries. The United States generated the largest amount, 29 percent of the world's nuclear electricity; France generated 17 percent, Japan 11 percent, and the former Soviet Union, 10 percent. These four regions accounted for two-thirds of the world's nuclear electricity generation in 1993. In the Third World, only four regions had nuclear-generating capacity in excess of 1,000 MWe: South Korea, Taiwan, People's Republic of China, and India. Of these four, nuclear power made a significant contribution (over 5 percent) to the installed electricity capacity of only the first two, both of which are relatively small in terms of their share of the global population and electricity use.[241]

The U.S. and global energy systems pose severe threats to the world's well-being that derive, albeit in different ways, from both nuclear energy and large-scale fossil fuel use. Nuclear energy carries with it the risk of catastrophic accidents like Chernobyl and proliferation liabilities arising from rapidly growing, widely distributed stocks of weapons-usable materials in the commercial sector. Fossil fuel use anywhere near current levels poses grave risks from the continued build-up of greenhouse gases.

The structure of the world's energy economy has left us with no really sound options in the short and medium term. The problem of policy is how to reduce the duration of poor choices and reduce the risk of adverse consequences of our actions today, while creating an environmentally sound and secure energy system for the future. This cannot be accomplished unless there is a basic re-evaluation of the technical, economic, and institutional premises of energy policy both in the United States and globally. However, so far as the Third World is concerned, where there is the greatest need for new commercial energy facilities, the share of nuclear electricity is very low in all but two relatively small areas: South Korea and Taiwan.

In countries that do not now have nuclear power, it would be prudent to by-pass the nuclear option altogether with sounder energy choices, given the problems associated with nuclear power in relation to proliferation, safety, and waste. But doing this will be difficult in practice if the industrialized countries, notably the United States, France, Japan, Russia, and Britain, insist on sticking to the nuclear

---

[241]Global nuclear electricity data are from NRC 1995.

course, while trying to deny it to others. Indeed, that would be a violation of the treaty commitment under the NPT of the nuclear weapons states to almost all non-nuclear weapons states, which are signatories in good standing of the NPT. The NPT went into effect in 1970 and was indefinitely extended in 1995.

One of the central problems with nuclear power is the potential transfer of the technology and skills to the production of nuclear weapons. This issue, like others in non-proliferation, cannot be resolved until the nuclear weapons states commit themselves firmly and clearly to a course of nuclear disarmament, which they show no sign of doing. On the contrary, despite their commitments under Article VI of the NPT to negotiate disarmament in good faith, they show every indication of continuing to hold on to large nuclear arsenals.[242] Further, as can be seen from the amounts of plutonium given in Table 6 (in Chapter 6), the continued reprocessing of commercial spent nuclear fuel is rapidly increasing the global inventory of separated plutonium. This is an especially urgent danger, given the precarious internal economic and political situation in the former Soviet Union after the end of the Cold War. Finally, as we have noted, the lack of any socially acceptable approach for addressing the problem of long-term management of spent nuclear fuel is exacerbating the proliferation problems associated with nuclear power.

When these proliferation concerns are joined to the fundamental unresolved safety issues associated with nuclear power, there is a strong case to be made against new nuclear power plants and for phasing out existing ones. However, this must be done in a manner compatible with the security of electricity supply and the reduction of the emissions of carbon dioxide from fossil fuel use. There is also an economic case to be made against nuclear power in the context of reducing greenhouse gas emissions. In the rest of this chapter, we briefly address some technical and policy issues that are related to the broad long-term goals of phasing out nuclear power and reducing carbon dioxide emissions, in a manner compatible with addressing the world's energy needs. We will then compare natural gas combined-cycle systems to nuclear power plants for their effect on carbon dioxide emission reductions. Finally, we will discuss the Kyoto Protocol.

---

[242]Makhijani and Zerriffi 1996.

Energy Concepts

The idea that the *quantity* of energy use is connected to economic growth, to levels of material consumption, and hence to human well-being is strongly entrenched. A detailed critique of these assumptions is beyond the scope of this book, but we will address them briefly. As a starting point, let us list some particulars that are widely acknowledged:

- A substantial majority of people in the world today do not have the basic amenities to live a modestly comfortable, secure life, and a large proportion of them live in terrible poverty.

- The population of the world is about six billion people and is expected to increase to roughly ten billion in the next several decades.

- The present manner of resource exploitation, use, and disposal is unsustainable and destructive to the environment and must be changed in order to meet the needs of the world's population.

- The stock of amenities, in terms of better homes, water supplies, transportation systems, health care systems, appliances, and the like, needs to be greatly increased to meet the needs of a majority of the world's population.

- In order to accomplish the increase in material well-being for the majority of the world's population that does not enjoy it today, two things are necessary from a physical point of view:

  1. the total *stock* of material goods in the world needs to increase considerably; and

  2. the flow of *energy* services, as defined by the accomplishment of tasks such as home heating, cooking, lighting, refrigeration, etc., needs to increase greatly, especially in Third World countries.

In the past two decades, analysts of energy and materials policies have recognized that amenities can be produced with decreasing quantities of materials through understanding of the economy of use of materials in natural systems. Where steel was the metaphor for strength of materials in the nineteenth century and much of the twentieth, the silk of spider webs is becoming the exemplar of materials engineering for the next. Further, the changes in materials use through recycling in the past decade indicate that the stock of material amenities can be largely disconnected from the despoliation of the natural environment so far as raw materials extraction is concerned.

A continual flow of useful energy through the economic system is required even to maintain the existing stock of amenities, not to speak of increasing it. In the past two decades, the distinction between the use of primary fuels and the delivery of the energy services that those fuels make possible has become widely appreciated. Experience and detailed analysis show that great increases in energy services are possible without significant increases in primary energy consumption due to the poor efficiency of present patterns of energy use.[243] While this is now well established, an example will illustrate the magnitude of the problem of inefficient energy use even in the most technologically advanced countries.

The most efficient gas heating furnaces for home central heating available in the United States are advertised as having efficiencies of over 90 percent. But this is a misleading claim, for it is the mere ratio of the amount of energy output to the energy input, without regard to the quality (or temperature) of that energy. If we take an energy source, such as oil, gas, or electricity, that has a high quality—that is, a high flame temperature—and degrade it to a few tens of degrees above room temperature, we have thrown away most of the capacity of these energy sources to perform mechanical work.

In a whole range of applications, involving almost all energy use from lighting to mechanical drives to heating and cooling, energy use is very inefficient. It is within technical possibility to reduce primary energy consumption by *severalfold* in the United States without changing substantially other major aspects of the economy. Much of this is within the range of available technology, while technological development is required to achieve the rest. For instance, within the next few

---

[243]See, for example, Goldemberg et al. 1988. In 1974, the Ford Foundation Energy Policy Project created a Zero Energy Growth scenario to show that primary energy use could level off by 1985 without affecting economic growth. Actual energy use in 1985 was lower than that projected in the Ford Foundation Zero Energy Growth scenario. (Ford Foundation 1974.)

decades it should be possible to increase the efficiency of residential and commercial heating and air-conditioning by severalfold through the use of cogeneration, fuel cells, and other technologies.

Energy for meeting the present level of amenities can be further reduced by dramatically changing the relationship of raw materials use to the production of goods and services. Put another way, primary energy use can be largely decoupled from the growth of amenities in the world. In countries such as the United States with a high level of primary energy use, positive economic growth is possible with reduction of primary energy use, since the room for efficiency improvement is great and the underlying quantity of energy use is very large. How these amenities are to actually reach those who need them is, of course, a difficult question beyond the scope of this work. But we note here that this is primarily a question of distributive equity rather than technical possibility.

Energy use in Eastern Europe and the former Soviet Union is far more inefficient than in the United States, western Europe, and Japan. In Russia, for instance, thermostats to control central heat are rare, so that those who live near the front end of district heating system pipelines are often too hot and open their windows for control of indoor temperature, while those at the tail end of the steam pipeline freeze.

In the Third World, energy use is far larger than shown in most compilations of data. Traditional fuels, like wood and cow dung, are common fuels and the efficiency of their use is very low.[244] For instance, the two tons of wood that are typically used per person to make charcoal for cooking in urban Kenya could be used to generate electricity to supply the cooking energy requirements for up to 20 people. But the investment priorities in Kenya, as in most of the Third World, are not oriented to meeting the needs of the urban and rural poor even for clean water supply, much less energy for cooking. Thus, the problem of a sufficient supply of useful energy is not so much a question of the availability of primary energy source. It is rather that the poor lack sufficient power and organization to channel investment priorities into such areas as clean water supply and efficient energy systems.

Finally, we need to establish criteria by which to judge a sustainable energy system. The following seven criteria, if they are met si-

---

[244]Goldemberg et al. 1988; and Makhijani and Poole 1975.

multaneously, could result in an environmentally sustainable and economically viable system.

1. It must be reliable.

2. Its cost should be reasonable.

3. It should not produce routine severe pollution.

4. It should be possible to almost wholly confine the environmental and security costs of the energy system to the generations benefiting from it. In other words, the system should be amenable to cost internalization.

5. It should be capable of sustaining reasonable levels of energy services to eight to ten billion people (the projected population of the world in the next century).

6. Its core functions should be resilient to supply, transportation, transmission, and economic shocks.

7. It should not add matter or energy flows to natural systems to an extent comparable to pre-existing natural levels, or fluctuations in those levels.

Renewable Energy Sources

Even with large improvements in energy efficiency, it will still be necessary in the coming decades to increase global primary energy supply, even if in the long term (say, after a hundred years or more), global energy consumption might be lower than present levels with far higher levels of energy services being provided with dramatically new and more efficient ways of meeting human needs.

The current world and U.S. energy supply mixes threaten the global environment and global security. To remedy this problem, the energy supply goals for renewable energy sources for the next several decades must become far more ambitious than they have been in any country so far. It is possible for renewable energy sources to constitute a rapidly increasing fraction of energy supply throughout the world. For instance, Johansson et al. estimate that, with the appropriate policies, renewable energy sources could supply about 60 percent of the world's electricity generation by the year 2025 and remain at that level thereafter.[245] Their scenario includes more than a doubling of overall electricity generation between 1985 and 2025, accompanied by considerable increases in efficiency of electricity use. They include small increases in global nuclear electricity generation in the coming decades. They also assume considerable decreases in coal-fired electricity generation to accommodate the need for reducing carbon dioxide emissions.[246] Since nuclear electricity would contribute *less than 6 percent* of the electricity supply by 2050 in their scenario (less than 9 percent by 2025), it should be possible with a modest adjustment in the pattern of investment to phase it out globally in the coming decades. Further, Johansson et al. consider only "commercial" energy sources and ignore the current traditional use of biomass fuels in the world. This source, in thermal energy terms, is about *two times larger* than the nuclear energy generation in their scenario for the year 2025. If the policies to increase energy efficiency and cogeneration are adopted, and if biomass is used far more efficiently in the Third World than it is today, nuclear power plants could be phased out while carbon dioxide emissions would also decline more than projected in the Johansson et al. scenario.

Table 12 on the following page shows a possible projection of global energy use without the use of nuclear energy, based on Johansson et al. We have adjusted the scenario of Johansson et al. to phase out nuclear power and to reflect the conversion by the year 2050 of all of traditional fuel use to modern modes by gasification and/or electricity generation, which are conducive to high efficiency. Of this, three-quarters is assumed to be electricity. The figure for the year 2025 would be half of the goal for the year 2050.

By the year 2050, the carbon dioxide emissions implicit in Table 12 would not only be below today's levels, but also below the levels assumed by Johansson et al. in their scenario, which includes continued, constant use of nuclear electricity at a level comparable to levels in the 1990s. This is because we have factored in potential energy efficiency gains that can be made in the use of traditional fuels. It is not necessary to include the entire amount of traditional fuel in the scenario to eliminate nuclear power. It would be possible to simply eliminate half of the use of traditional fuels, if the use of the other half

---

[245]Johannson et al. 1993, p. 2 and Table 5, p. 1104, Appendix A.

[246]Johansson et al. 1993, Appendix A.

**Table 12: Renewable Intensive Global Electricity Supply
Projection
(Billion Kilowatt Hours per Year)**

Fuel Source	Year: 1985	Year: 2025	Year: 2050
Coal	3,782	2,300	1,000
Oil	1,042	0	0
Natural Gas	1,099	4,700	8,000
Nuclear	1,399	300	0
Hydroelectricity	1,880	3,800	4,800
Intermittent renew- ables (solar plus wind)	0	4,700	9,600
New biomass sources	0	3,900	5,800
Converted traditional biomass	0	1,500	3,000
Geothermal	15	200	200
Other	21	0	0
Total	9,238	21,400	32,400

Source: Adapted from Johansson et al., eds. 1993, Table 5, Appendix A, p. 1104. Figures for 2025 and 2050 taken from Johansson et al. have been rounded.

Notes: 1. Assumes use of traditional fuels in 1985 = 40,000 petajoules. Amount of nuclear electricity offset by converting traditional biomass to electricity in 2025 and 2050 = 1,800 billion kilowatt hours. Amount of coal generation offset by traditional biomass in the year 2050 = 1,000 billion kilowatt hours. Amount of natural gas generation offset in the year 2050 = 200 billion kilowatt hours.

2. Nuclear phase-out assumed to be complete by 2030, which would mean that most plants would be retired after their licensed lifetimes, but that some retirements (such as those of relatively unsafe plants in the former Soviet Union) would be accelerated, reflecting events of the 1990s. It is possible with modest adjustments in the assumptions to come up with slower or faster retirement schedules. These would be speculative at this stage, since the rate of use of renewables is far from that required to bring about the changes assumed in the table.

is via modern modes. This may be desirable from the point of view of using land for agricultural rather than energy purposes, for instance. Second, none of the scenarios we have discussed here take into account the potential for improving the efficiency of the use of land and biomass resources for draft animals in agriculture.[247] Finally, the pace of phase-out of nuclear power could be very different from the one assumed below. The assumption in Table 12 is that some countries, most notably France, will find it difficult to phase out nuclear power rapidly even if a policy in that direction were adopted because it supplies a large proportion of their electricity (78 percent in the case of France in 1993).

The scenario in Table 12 assumes rather heavy use of biomass fuels, which may be limited by the need to use land and biomass resources for other purposes. It is possible to construct other scenarios with greater emphasis on wind energy and natural gas. Wind energy is concentrated in some regions, and hence it can provide large amounts of electricity to centralized grids. The world's natural gas resource base is quite large and dispersed more widely than coal. Natural gas can be used to generate electricity more efficiently and at considerably lower costs than nuclear power plants. Finally, sequestration of carbon dioxide so that it is not emitted to the atmosphere and therefore does not contribute to greenhouse gas emissions can extend the use of coal. The Intergovernmental Panel on Climate Change estimates that sequestration can reduce carbon dioxide emissions from natural gas plants by a factor five for an overall cost of $1,420 per kilowatt electrical, which is lower than the cost of a nuclear power plant. The capital cost of a coal-fired plant that would have ten times lower greenhouse gas emissions compared to present plants with sequestration is estimated at about $3,000 per kilowatt electrical. Costs are in 1990 dollars.[248] This last figure is comparable to recent costs of nuclear power plants in the United States. Finally, it is noteworthy that creating an infrastructure for use of natural gas in applications, such as transportation, is compatible with the use of hydrogen from a variety of renewable energy table sources in the long term. Hydrogen is potentially the soundest energy source from an environmental point of view, provided it is generated from renewable energy sources and used appropriately.

---

[247]For discussions on this subject, see Makhijani and Poole 1975; and Makhijani 1990.

[248]IPCC 1996, Table 19-1, p. 593.

Wind turbines at Tehachapi, California. With appropriate policy decisions, such renewable energy sources as wind can constitute a rapidly increasing fraction of energy supply throughout the world. (U.S. Department of Energy)

Integrating Renewables with Energy Efficiency in the Electricity Sector[249]

It is now well over 20 years since the first oil crisis broke upon an unready United States. The oil-shock and the higher energy prices that followed abruptly changed historical patterns of energy use. The ratio of the growth rate of energy use to the growth rate of GDP prior to 1973 had been about 1:1. By the mid-1980s, energy use was at about the 1973 level even as GDP had increased by about 25 percent. In other words, the ratio of energy growth to GDP growth went to zero, something many economists of the 1960s had thought practically impossible. The Gross Domestic Product per unit of energy use increased from $41 per gigajoule in 1970, to $55 per gigajoule in 1985, to $58 per gigajoule in 1993 (GDP figures in 1987 dollars). The growth rate of electricity declined, but the growth continued. [250] Energy use has climbed since the mid-1980s; the rate of electricity growth has also been creeping upwards. Both of these changes correspond to declining producer prices for energy and a slower pace in energy efficiency gains.

The discussion below is specific to the United States, although most of its elements could also be applied elsewhere with appropriate adjustment to reflect local circumstances.

1. Obstacles to Efficiency

The 1970s and 1980s have seen enormous technological changes in electronics, in some generation technologies, in technologies relating to energy efficiency, and in the application of computers to the control and management of electric power systems. This has created the potential for technical efficiency gains and for competition in electric power in ways not possible two decades ago. But despite the technological potential, the overall picture in the electricity sector, and in the energy sector as a whole, remains a disquieting one.

There is currently no framework for ensuring reasonably priced electricity for all consumers or for creating a flexible system that could

---

[249]This section is an edited, updated, and condensed version of Makhijani 1995a.

[250]Figures here and below calculated from energy, electricity, and GDP data in the 1995 edition of the *Statistical Abstract of the United States*.

also meet challenges such as the large-scale use of electric cars and reductions in carbon dioxide emissions simultaneously and rapidly.

Despite all the demand management investments[251] of the last decade-and-a-half, there has been no overall improvement in the use of electricity per unit of economic output since the 1970s. The ratio of GDP (in 1987 dollars) to electricity was $1.88 per kWhe in 1970, $1.65 per kWhe in 1980, and $1.78 per kWhe in 1993. This is in large measure due to the increasing proportion of energy use that is in the form of electricity, which has driven some of the gains in energy efficiency. But given that there are fewer energy intensive industries in the United States and that the economy has moved toward the service sector, the relatively static figure of output per unit of electricity also indicates inadequate efforts to improve efficiency of electricity use per unit of economic output despite economic opportunities for efficiency improvements. This gap between potential and reality is due to fundamental obstacles:

- *Inefficiencies in the electricity market*: The market in electricity is inefficient and unpredictable primarily because the time scale for creation of new demand, such as building new buildings or factories (or canceling them), is much smaller than the time-scale of construction of new power plants, especially large central station power plants. No matter how sophisticated forecasting programs might get, they cannot make up for this basic dysfunctional feature in the present structure of electricity supply and demand.

- *Large stock of inefficient existing buildings and inadequate efficiency of new buildings and investments*: The difficulties of realizing potential efficiency increases in existing buildings are great and the cost of capturing them is often high. New structures are also generally far less efficient than they need to be.

- *Lack of a systems approach to achieve flexibility and reliability*:

---

[251]Demand management is the adjustment of electrical demand to allow for efficient use of electrical generation capacity. It is not restricted to investments in energy efficiency, but includes measures such as utilities remotely switching off air-conditioners for brief periods during peak usage hours.

Long lead times for centralized investments create unnecessary financial risks due to their inflexibility. The system of maintaining reliability by requiring a certain peak margin has also become costly because of the long lead time required to build new power plants. Peak margin, which is the amount of spare capacity needed at the moment of greatest electricity demand time to keep power supply reliable, must be maintained above the minimum required for reliability throughout the period of construction. If lead times are long and capacity additions are made in large increments, then the average amount of idle capacity increases, raising costs. While purely centralized systems generally provide higher reliability per unit cost than purely decentralized systems, distributed electricity systems can be optimized to be better than both. Yet there are at present no consistent technical, institutional, and financial mechanisms for combining centralized and decentralized sources to achieve distributed electricity supply systems. This is evolving to some extent with the advent of deregulation of electricity generation and potential conversion of transmission networks into common carriers.

- *Institutional factors.* There a host of institutional factors that prevent wider use of available technological opportunities for energy efficiency improvements. Examples include lack of adequate information on the part of consumers; the fact that developers generally do not pay energy bills; the relatively low percentage of overall domestic, industrial, and commercial budgets represented by electricity; and, not least, the failure of policymakers to require energy prices to reflect environmental costs are fully as possible.

- *Lack of adequate government and utility procurement policies.* Robert H. Williams, a senior researcher at Princeton University, has noted that government research and development policies on developing new technologies, and bringing them to market, have not been very effective in the energy arena. He further observes that, while energy R&D is necessary, procurement is a much more effective tool for shaping the marketplace in the direction of technologies that are clearly desirable.[252]

The most important of these structural problems is the lack of a rational market for electricity. The second most important factor, but perhaps the most important obstacle in the short-run, is the lack of an adequate procurement policies on the part of governments and utilities.

An electricity market must, of course, have the basic feature of markets—that is, supply and demand must be closely coupled. But a rational market for an essential commodity that has a huge impact on the global economy and environment must also take into account a number of other concerns:

- It must maintain at least the present level of reliability of supply for all consumers who are on a common grid.

- It must be able to provide everyone with reasonably priced electricity.

- It must be flexible enough to respond to the needs of environmental protection. In particular, it must have the flexibility to reduce the use of coal and oil and increase greatly the efficiency of primary energy use in order to accommodate likely requirements for reduction of greenhouse gas emissions in the coming decades. It must also be compatible with phasing out nuclear power.

- It must be flexible enough to respond to changing technology.

In sum, creating a rational market in electricity means not only that the system will generate returns on investment, but also that it be environmentally sustainable, socially equitable, and technologically responsive. The single most important key to this is to create a financing structure for the electricity sector that joins investments in supply and demand so as to create a close coupling between them.

2. A Proposal for a Financing Structure

Electricity supply and demand could be linked by the requirement that new residential, commercial, and industrial developments contri-

---

[252]Williams 1993, p. 4-40.

bute to an electricity system capital fund an amount corresponding to the proposed electricity demand they plan to make on the grid. Currently developers have little incentive to optimize investments relating to the energy characteristics of their structures because they either do not pay the energy bills or simply collect from tenants as part of rent. Requiring developers to put money into a capital fund would encourage them to consider investments in energy efficiency, cogeneration, and renewable energy sources. This would reduce the amount that they would have to pay into the fund. The net effect of such a policy would be to reduce the electricity to GDP growth rate ratio because the requirement of putting up the money for a capital fund would spur investments in energy efficiency that would otherwise not be made.

The creation of a capital fund would allow flexible and rapid response to changing market conditions, new requirements for environmental protection, and the introduction of new energy generation or utilization technologies. In fact, the existence of such a fund is likely to spur complementary investments in research and development and introduction of goods that will meet the needs for a better electricity system.

The capital fund could be set up as a public corporation, regulated by state public utility commissions. It would be a source of capital for utilities and others to use for efficiency and supply investments in order to meet new demands on the system. The fund could be invested in large measure in the areas of greatest return. For instance, if it is most efficient to invest in improving the efficiency of existing buildings, then there would be a source of capital available to do so. The capital fund could also be used to finance government and utility procurement of electricity generated from renewable energy sources.

3. Improving Efficiency in Existing Large Buildings

One of the most promising large sources of new investment in increasing supply and greatly improving efficiency lies in large existing buildings—that is, buildings greater than a couple of hundred kilowatts peak demand. Converting existing large buildings, using natural gas or oil heating systems and electric compressor-driven air-conditioning to cogeneration, would "mine" existing buildings for natural gas by increasing efficiency greatly. Waste heat would be used for both heating and air conditioning (via absorption air-conditioning systems). This would have the added advantage of eliminating ozone-depleting compounds now used in air-conditioning systems. Installation of cogeneration could be used to optimize the other demands on the electricity system, notably lighting and motors, since investments to increase their efficiency would mean either a smaller size of cogeneration plant or a larger amount of electricity available for sale to the utility. Thus the economies of improved efficiency in lighting, motors, and other systems would be more easily realizable than they are without the installation of cogeneration systems.

4. Improving Efficiency in Existing Small Buildings

The ability to convert existing small commercial and residential buildings to some type of cogeneration system would vastly increase system flexibility, allowing for simultaneous increases in supply and efficiency. At present, most systems below a couple of hundred kilowatts are too expensive to justify retrofits on their own. But limited retrofits could be justified to meet unanticipated sudden increases in demand, such as that due to electric cars, for instance. Experimental conversions of a number of existing residential and small commercial structures to various combinations of natural-gas-based fuel cells, solar photovoltaics, and the like, coupled with highly efficient end-use technologies, are needed to enable study of system optimization from the points of view of efficiency, reliability, emissions, and cost. Such systems would be connected to the grid for reliability and optimization.

5. Other Measures

Electricity systems impinge on vital security and environmental issues. Therefore, it is appropriate for governments to adopt particular procurement policies to create a demand for desirable technologies. For the same reason, public utility commissions can require a modest amount of investment in such technologies on the part of utilities so that they may become competitive with other sources of electricity on a faster schedule and can be better integrated into the overall electricity system. A similar principle already has widespread acceptance in the United States. Public utility commissions have been requiring utilities to make investments in energy efficiency at a modest levels for many years. Technologies that are commercial (such as wind power in some areas) or close to commercial (such as solar-thermal generation in other areas), or on a rapid track to becoming commercial (photovoltaic or solar cells) can be procured in this way. Competitive bidding for fixed amounts of such electrical capacity would create a faster and surer track to their commercialization and adoption.[253] Similarly, the federal government could also acquire renewable technologies for a part of its energy needs.

Finally, a hierarchy of "just-in-time" investments can be developed to reduce the need for risky, long lead time, centralized power plants. Investments with short lead times can reduce the average capacity margin corresponding to a given minimum capacity margin required for reliability. Even a modest reduction would save large amounts of money by making new power plant construction unnecessary.

Combined-Cycle Plants, Nuclear Power Plants, and CO_2 Emissions

Proponents of nuclear energy suggest that the problem of greenhouse gases can be solved by nuclear power because nuclear reactors do not emit carbon dioxide into the atmosphere. However, the high costs of nuclear power make it unsuitable even if the other criteria are set aside.

Pro-nucler arguments are wrong even on narrow economic grounds because the expense of nuclear power would actually preempt investments in technologies more appropriate for reducing carbon dioxide emissions. Table 13 below shows that natural gas combined-cycle plants are not only more economical than nuclear power plants, but that for a given sum of money (in present value terms) the reduction in CO_2 emissions is considerably greater with combined-cycle plants than with nuclear power plants, assuming that these plants are used to replace existing coal-fired generation—a leading source of CO_2 emissions.

This may seem paradoxical at first, since natural burning produces CO_2 emissions and, to a first approximation, nuclear power plants do not.[254] But combined-cycle plants are far more efficient than the aver-

---

[253]Williams 1993, p. 4-40.

[254]All power plants, and indeed all investments, require a front-end energy investment. In this sense, building nuclear power plants, combined-cycle plants, solar energy systems, or wind turbines all contribute to some greenhouse gas emissions. But they reduce these emissions over the life of operation if we replace coal with any of them. Most nuclear power plants use enrich uranium fuel. Uranium enrichment takes a great deal of energy, but it is far less than the nuclear power plant output. When doing

Table 13: Comparison of Carbon Dioxide Reductions: Natural Gas Combined-Cycle versus Nuclear Power Stations

Power System	Capital Cost ($/Kw)	Interest + Depreciation (¢/kWhe[1])	Natural Gas Price ($/million Btu[2])	Fuel Cost (¢/kWhe)	Non-fuel O&M (¢/kWhe[3])	Total Cost (¢/kWhe)	Total CO_2 Reduction after 30 Years (kg C[4])	Carbon Reduction Ratio (Gas/ Nuclear)
Combined-Cycle (CC)[5]								
Case 1 CC	500	0.76	150	1.02	0.48	2.26	9.97×10^{10}	1.37 (Case 1)
Case 2 CC	500	0.76	250	1.71	0.48	2.95	1.02×10^{11}	1.40 (Case 2)
Case 3 CC	500	0.76	400	2.73	0.48	3.97	1.09×10^{11}	1.50 (Case 3)
Nuclear[6]								
Case 1 Nuc	1500	2.28	–	0.6	1.7	4.58	7.29×10^{10}	–
Case 2 Nuc	2500	3.81	–	0.6	1.7	6.11	7.29×10^{10}	–
Case 3 Nuc[7]	4000	6.09	–	0.7	2.0	8.79	7.29×10^{10}	–

Based on the following sources: For nuclear plant costs (Cases 2 and 3): Cohn 1997, pp. 106 and 155; NRC 1997, Tables 6 and 7. For gas costs: the U.S. Energy Information Adminstration Web page. For combined-cycle power plant costs: Todd and Stoll 1997; and Komanoff, Brailove, and Wallach 1997, p. 39.

Notes:

1. Interest and depreciation assumed to be 10 percent in all cases. Capacity factor assumed to be 75 percent in all cases.

2. Btu stands for British thermal units. One Btu = about 1,055 joules. One

---

calculations to a first approximation, it is reasonable to ignore front-end energy investment costs for all systems and to assume that nuclear power plants emit no CO_2 even though this is not precisely correct.

kWhe (kilowatt-hour electrical) = 3.6 million joules = 3,413 Btu.

3. Non-fuel nuclear costs include 0.2 cents per kWhe for waste disposal and decommissioning, except in the worst case (Case 3) where this cost is taken to be 0.5 cents per kWhe. (See Cohn, p. 155.)

4. The CO_2 emissions avoided are calculated on the assumption that both types of power plants would displace existing coal-fired power plants emitting 0.37 kilograms (carbon basis) per kWhe. For nuclear, the avoided emissions would therefore be 0.37 kg, to a first approximation. For combined-cycle with 50 percent efficiency, the figure is about 0.25 kg per kWhe (emissions from the coal-fired power plant less the emissions from the combined-cycle plant). The avoided CO_2 emissions figures for combined-cycle plants are likely to be increased for plants installed a few years hence because the efficiency of these plants is increasing.

5. Efficiency of the combined-cycle plant is assumed to be 50 percent. Higher efficiencies, approaching 60 percent, are expected in the next few years. We have assumed a natural gas fuel value of 1,000 Btu per cubic foot in these calculations. (Nuclear power plant thermal efficiency is about 33 percent. The exact figure does not affect power costs substantially, since fuel costs are a small fraction of total costs.)

6. Nuclear costs do not include any reprocessing and plutonium management costs.

7. The worst-case capital cost of nuclear (Case 3) was typical of U.S. costs for plants coming on line after 1983, but with far higher capacity factor than was typical of the 1980s in the U.S. The best-case nuclear capital cost (Case 1) is that reported by the media for sales of Russian VVER-1000 reactors to China.

age coal-fired power plant and natural gas burning results in only half the CO_2 emissions that coal does per unit of heat energy produced. Hence, even though a unit of electricity produced by nuclear reduces CO_2 emissions more than a unit of energy produced by natural gas combined-cycle plants, the reduction of emissions for a given sum of money is considerably greater for combined-cycle plants.

Combined-cycle plants use a fuel, such as natural gas, in a two-step electricity-generation system (see Figure 9 on the following page). First, the natural gas drives a gas turbine and a generator. Then the hot

Figure 9: Combined-Cycle Plant

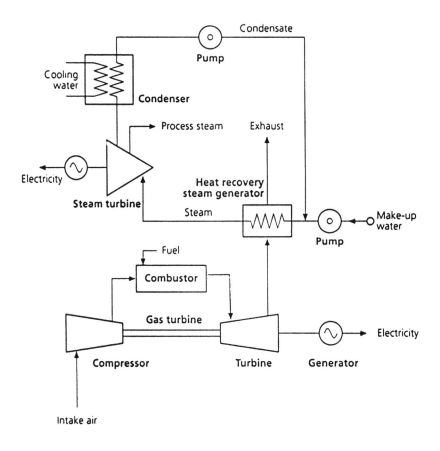

Granted with permission from *Renewable Energy: Sources for Fuels and Electricity*, Thomas B. Johansson, Henry Kelly, Amulya K.N. Reddy, and Robert H. Williams, eds., Fig. 9, Chap. 17, p. 740. © 1993 Island Press, published by Island Press, Washington, DC and Covelo, CA.

exhaust gases from the turbine are used to raise steam, which drives a steam turbine. The efficiency of such a system available commerically today is about 50 percent.

Note that China, the main prospective customer for new nuclear power plants, is unlikely to have the highest costs of combined-cycle plants because it would use piped gas (from its own onshore and off-shore fields as well as Central Asia) and not liquid natural gas (on which Case 3 costs are based).

This comparison excludes pessimistic scenarios for nuclear power plant costs, which would be substantially higher than the highest nuclear costs given in Table 13 (page 196). Each cent per kWhe difference in costs works out to about $66 million per year in additional electricity costs for nuclear power plants (1,000 MW size). This works out to a present value over a 30-year period (at an annual discount rate of 4 percent) of $1.15 billion for every cent per kWhe difference in electricity costs. (Future costs are discounted, since a dollar saved at a future time is worth less than a dollar in hand today.) Using these figures, one can compare a strategy of using nuclear power plants to displace existing coal-fired power plants with one of using combined-cycle power plants. In the table, we have compared the various cases for combined-cycle versus nuclear: low cost versus low cost, medium versus medium, and high versus high. For a typical case, building combined-cycle plants would result in a reduction of about 40 percent more CO_2 than could be achieved with nuclear (comparison of Case 2 combined-cycle with the corresponding nuclear power plant). This gain can be expected to increase since efficiencies of combined-cycle plants are increasing.

One could also use the capital cost savings achieved by building combined-cycle plants instead of nuclear to develop and promote solar and wind technologies and to increase energy efficiency. The avoided CO_2 emissions in such cases would vary depending on the sites for the power plants or the specific technologies chosen to increase efficiency. If combined-cycle plants were used to retire half the coal-fired power stations in the world, an overall annual global carbon dioxide emissions reduction of about 15 percent could be achieved.

During the 1970s, there was concern that natural gas was a very scarce resource, but it was not well founded. Gas is a widely available resource, and does not carry the proliferation risks of nuclear power. Our approach is not premised on use of natural gas into the indefinite future, but only on its use in high-efficiency applications over the next several decades. This use of natural gas as a transition fuel is a sound economic and environmental strategy. During that time we expect—with appropriate action on the part of governments, corporations, and consumers—that renewable energy sources will take over most of the energy supply in a vastly more efficient economy.

World reserves of natural gas have been steadily rising, and now stand at about 75 years of consumption at 1995 levels (corresponding to reserves of about 5.2×10^{21} joules in reserves, and an annual utilization of about 7×10^{19} joules). Global gas reserves have been steadily increasing, despite increasing consumption.[255]

Coal-fired power stations are located in many parts of the world, including western Europe, the United States, the former Soviet Union, China, India, and eastern Europe. While it is unlikely to be economically feasible to immediately replace coal-fired plants with combined-cycle plants, it is possible to phase out coal-fired plants and replace them over time. In some areas, wind capacity on land and off-shore would also provide an effective and economical offset for CO_2 emissions.

One drawback to increased use of natural gas is that natural gas pipelines add to methane emissions due to small leaks in the pipelines. One estimate of such leaks is 0.8 percent of natural gas use. Since methane is a far more powerful greenhouse gas than CO_2, it is necessary to offset these emissions in order to maximize the greenhouse gas reductions that can be obtained from natural gas use. Such offsets can be obtained by relatively simple measures, such as building biogas plants at feedlots or recovery of methane gas emitted from landfills (now a significant pollutant in many areas) for use as a fuel. Landfill gas is used on a limited basis in many places to produce electricity or fuel for heating. For instance, landfill gas from the Fresh Kills landfill, where the municipal waste from New York City is dumped, provides heating fuel for 14,000 homes.[256]

The Kyoto Protocol

From December 1 to 11, 1997, the Third Conference of the Parties ("COP-3") to the United Nations Framework Convention on Climate Change was held in Kyoto, Japan, and included over 10,000 participants from governments, intergovernmental organizations, non-gov-

---

[255]U.S. Energy Information Administration Web page.

[256]Vivian Toy, "Sealing Mount Garbage: Closing Staten Island's Fresh Kills Dump Is an Operation of Staggering Complexity," *The New York Times,* December 21, 1997.

ernmental organizations, and the press. In the Kyoto Protocol, adopted by 171 countries at the conference, structures were put into place to reduce six major greenhouse gases: carbon dioxide (CO_2), methane (CH_4), nitrous oxide (N_2O), hydrofluorocarbons (HFCs), perfluorocarbons (PFCs), and sulphur hexafluoride (SF_6). Major provisions of the Kyoto Protocol are below. (Note: "Annex I parties" are those countries included in Annex I to the United Nations Framework Convention on Climate Change, adopted in New York on 9 May 1992. "Annex B parties" refers to those countries included in Annex B in the Kyoto Protocol. See Table 14 on the following page.)

1. Major Provisions of the Kyoto Protocol

- *Greenhouse gas emissions reduced to 5 percent below 1990 levels*: Article 3 of the Protocol reflects the parties' commitments to reduce overall emissions of greenhouse gases by 5 percent below 1990 levels between 2008 and 2012. The 5 percent target is not a global target but applies as an overall target to a list of countries in Annex B of the Protocol. Some countries, including the U.S., Canada, European Union countries, and Japan, will have to reduce emissions up to 8 percent. Some on this list, including Australia and Iceland, will be allowed to increase emissions by varying amounts up to 10 percent. There are no limits for "developing" countries, including China, India, Brazil, Mexico, Indonesia, Nigeria, etc., where per capita consumption of fossil fuels is still relatively low.

- *Emissions trading*: Article 16 *bis* states that: "The Parties included in Annex B may participate in emissions trading for the purposes of fulfilling their commitments under Article 3 of this Protocol. Any such trading shall be supplemental to domestic actions for the purpose of meeting quantified emission limitation and reduction commitments under that Article."

    Article 16 *bis* was a late addition to the Kyoto Protocol, and a subject of contentious debate. It allows emissions trading in principle. Specific rules were discussed at the Fourth Conference of the Parties ("COP-4"), held November 2-13, 1998, in Buenos Aires, Argentina. They remain to be worked out.

Table 14: Annex I and Annex B Parties
(all are both Annex I and Annex B, unless otherwise noted)

Country	Kyoto Protocol Emissions Limit as Percentage of Base Year (1990)	Country	Kyoto Protocol Emissions Limit as Percentage of Base Year (1990)
Australia	108	Lithuania	92
Austria	92	Luxembourg	92
Belarus (Annex I only)	–	Monaco (Annex B only)	92
Belgium	92	Netherlands	92
Bulgaria	92	New Zealand	100
Canada	94	Norway	101
Croatia (Annex B only)	95	Poland	94
Czech Republic	92	Portugal	92
Denmark	92	Romania	92
Estonia	92	Russian Federation	100
European Community	92	Slovakia	92
Finland	92	Slovenia (Annex B only)	92
France	92	Spain	92
Germany	92	Sweden	92
Greece	92	Switzerland	92
Hungary	94	Turkey (Annex I only)	–
Iceland	110	Ukraine	100
Ireland	92	United Kingdom	92
Italy	92	United States of America	93
Japan	94		
Latvia	92		
Liechtenstein	92		

- *"Clean Development Mechanism"*: Article 12 defines a clean development mechanism, the purpose of which is to assist developing countries to achieve "sustainable development." Annex I countries could count reductions in greenhouse gases achieved in this way against their own targets.

- *Joint Implementation*: Article 6 states that: "for the purpose of meeting its commitments under Article 3, any Party included in Annex I may transfer to, or acquire from, any other such Party emission reduction units resulting from projects aimed at reducing anthropogenic emissions by sources or enhancing anthropogenic removals by sinks of greenhouse gases in any sector of the economy"

  While similar to the "clean development mechanism," joint implementation refers to the trading of emissions reduction units among Annex I parties (generally, industrialized countries), while the clean development mechanism allows Annex I parties to benefit (i.e., gain emission reduction units) from emissions reduction projects performed by corporations in non-Annex I countries.

2. Joint Implementation

Joint implementation, a key part of the provisions of the Kyoto Protocol, involves the trading of emissions between two parties. The idea is that if one party can reduce emissions more cheaply than another, or is already below allowable limits, then the party for whom it would be more expensive could simply purchase emissions reductions. This avoids the added expense involved when all polluters must reduce their own emissions. Thus, in theory, by relying on "market principles," society would achieve emission reduction targets at the lowest cost.

This theory has been tried out with some success in the United States for reducing sulfur dioxide (SO_2) emissions from industrial sources. These sources, such as many coal-burning power plants, are large emitters of SO_2, and their emissions are relatively well known. It has enabled industries that might otherwise face a shutdown to prolong their timetables for achieving compliance.

The following conditions appear to be required for emissions trading to be successful:

- The implicit price of a unit of pollution should be high enough to provide a substantial incentive to all polluters to reduce emissions.

- There should be a progressive tightening of targets toward the desired levels, so that the desired reductions are actually achieved in a reasonable time.

- The sources of emissions should be well characterized, so that the progress in emissions reductions can be measured with confidence. This is a key requirement, since without it, enforcement would be impossible and questionable schemes would flourish.

- Price negotiations should be between parties of comparable economic power, so that trading is equitable. These conditions were all met in the U.S. experience with SO_2 trading, and generally hold in the case of emissions from large industries within countries. They may also be roughly fulfilled when large industries in different countries negotiate across borders, although factors such as differences in currency convertibility and inequitable exchange rates must be taken into account.

In the case of CO_2 emissions regulated under the Kyoto Protocol, the units of account are countries themselves, so that domestic trading is not at issue. (Each country may, of course, opt to have CO_2 trading permits within its boundaries to achieve its Kyoto Protocol targets, but that is within the province of that country's government and not the Protocol itself.) Trading of emissions between large industries, such as power plants, located in most countries listed in Annex I (or Annex B) may be appropriate, provided the pricing arrangements are worked out. (See Table 14 above for list of Annex I and Annex B countries.) However, since there are many economically weak countries with weak currencies on the Annex I list (such as the former Soviet Union and eastern European countries), trading may become inequitable. Moreover, the pre-1990 records of emissions from large industrial plants in the former Soviet Union and eastern Europe are likely to be poor or incomplete in many cases. Finally, the relevance of these records for the next decade is highly questionable, given the huge changes that have taken place since 1990.

If trading between Annex I countries for the purposes of joint implementation appears to be problematic, it will be even more so between Annex I countries and developing countries. Besides the measurement and enforcement questions, the equity issue is particularly serious here. The CO_2 problem has been caused primarily by emissions from the industrialized countries. But emissions rights are being allocated on the basis of 1990 levels, giving the lion's share of the value of emissions credits that could be traded to those who created the problem. The countries with lowest fossil fuel consumption would hold the lowest emissions credits and hence derive the least benefit, although they have contributed least to the problem. This is a central reason that these countries did not agree to emission limits for themselves in the Kyoto Protocol. This keeps open until a later date (presumably the meeting in Argentina in November 1998), the question of what level of emissions trading rights developing countries will have.

If the emissions rights were on a per capita basis, as many people in the developing countries are demanding, then the feasibility of joint implementation would be considerably expanded, as would the economic benefits to be derived from it.

Proposals for joint implementation, involving sectors other than industry, pose additional problems. Examples include planting forests in developing countries and using agricultural residues in power plants to offset CO_2 emissions in industrialized countries. Such proposals are ill-suited to joint implementation, as they do not meet several of the conditions set forth above. First, the inequality of the negotiating parties in such arrangements is evident, and is exacerbated by the fact that upper classes in highly class-divided societies do the negotiating on behalf of farmers and the poor.

The technical issues are equally daunting. In the simplest instance, a tree would contribute to CO_2 reductions only during the growth period. After that there may be a net increase or decrease in CO_2, depending on the specific circumstances. Further, emissions of greenhouse gases, such as nitrous oxide and methane, would need to be taken into consideration.

Would replanting be required? How would one account for natural changes in the forest area over time? It is quite unclear how such complex processes would be factored into greenhouse gas accounting.

There is also the question of land. Developing countries do not have much idle productive land. Common land and partly forested land is often used by the poor for grazing draft animals, as a source of fuelwood and construction wood, and for other uses. Monetizing this land by making it a part of joint implementation projects could deprive millions of poor people access to basic resources, even though they did not create the present greenhouse gas problem. For in-depth discussion of CO_2 emmision trading, allocation of carbon-emission quotas, and related subjects, see Dubash 1994; and Kinzig and Kammen 1998.

Even many projects that appear attractive on the surface do not stand up to scrutiny. For instance, a project that might use bagasse (leftover organic matter after the juice is crushed out of sugarcane stalks) for generating electricity in India could have disastrous consequences of large numbers of people. Bagasse is already used for a variety of purposes, including electricity generation. The pressure to get more bagasse for such generating plants could wind up increasing sugarcane cultivation while displacing food crops. Further, many places in India use bagasse in traditional furnaces to make raw brown sugar, known as "gurdh." Joint implementation might kill such traditional industries, which employ large numbers of people, and create even more unemployment in already distressed rural areas. Or it may force gurdh manufacturers to use fuelwood instead of bagasse, possibly increasing CO_2 emissions. In sum, there are limited prospects for joint implementation and these could perhaps be pursued with some economic benefit to the global community. However, they would have to be carefully thought through in a way that has not yet been done to ensure that they are equitable, that their results are measurable, and that the poorest populations of the world are not adversely affected.

APPENDIX A:
BASICS OF NUCLEAR PHYSICS AND FISSION

Structure of the Atom

The atoms, of which every element of matter is composed, have a nucleus at the center and electrons whirling about this nucleus that can be visualized as planets circling around a sun, although it is impossible to locate them precisely within the atom. The nuclei of atoms are composed of protons having a positive electrical charge, and neutrons that are electrically neutral. Electrons are electrically negative and have a charge equal in magnitude to that of a proton.

The number of electrons in an atom is normally equal to the number of protons in the nucleus. As a result, atoms of elements are normally electrically neutral. The mass of an atom lies almost entirely in its nucleus, since protons and neutrons are far heavier than electrons.

Free neutrons are unstable particles which decay naturally into a proton and electron, with a half-life of about 12 minutes.

neutron ———> proton + electron + an anti-neutrino

However, it is remarkable that neutrons, when they exist together with protons in the nucleus of atoms, are stable. Protons are about 1,836 times heavier than electrons, and neutrons are about 1,838 times heavier than electrons. The energy balance in the decay of a neutron is achieved by the anti-neutrino, a massless, neutral particle that carries off surplus energy as the neutron decays. The nominal mass of an atom of an element is measured by the sum of the protons and neutrons in it. This integer is called the *mass number*. The nominal mass of an atom is not affected by the number of electrons, which are very light. Hence, the nominal mass based on the mass number approximates the actual atomic mass. The number of protons in the nucleus, which determines the chemical properties of an element, is called the *atomic number*.

207

Elements are arranged in ascending order of atomic number in an arrangement called the periodic table. The term derives from the tendency to periodicity of chemical properties deriving from arrangements of electrons in atoms.

Radioactive Decay

The nuclei of some elements are not stable. These nuclei are *radioactive*, in that they emit energy and particles, collectively called "radiation." All elements have at least some isotopes that are radioactive. All isotopes of heavy elements with mass numbers greater than 206 and atomic numbers greater than 83 are radioactive.

There are several ways in which unstable nuclei undergo radioactive decay:

- Alpha decay, which the emission of a helium-4 nucleus containing two protons and two neutrons. This is the least penetrating form of radiation. It is stopped by the dead layer of skin and so does no harm when outside the body. But it is the most damaging form of radiation when deposited inside the body.

- Beta decay, which the emission of an electron or a positron (a particle identical to an electron except that it has a positive electrical charge).

- Electron capture, which is the capture by the nucleus of an electron from among the ones whirling around it. In effect, the electron combines with a proton to yield a neutron.

- Spontaneous fission, which is the fission of a heavy element without input of any external particle or energy.

Often, there is still excess residual energy in the nucleus after the emission of a particle or after electron capture. Some of this residual energy after radioactive decay can be emitted in the form of high-frequency electromagnetic radiation, called gamma rays. Gamma rays are essentially like X-rays and are the most penetrating form of radiation.[257] It should be noted that the emission of gamma rays does not change the mass number or atomic number of the nucleus—that is, unlike radioactive decay by emission of particles, spontaneous fission, or electron capture, it does not cause the transmutation of the nucleus into another element.

Each quantum or unit of a gamma ray (or other electromagnetic energy) is called a *photon*. Gamma rays are like light, except that they are much higher frequency electromagnetic rays. Photon energy is directly proportional to the frequency of the electromagnetic radiation. Photons of gamma rays can damage living cells by splitting molecules apart or ionizing elements in them.

Many heavy nuclei emit an energetic alpha particle when they decay. For instance, uranium-238 decays into thorium-234 with a half-life of almost 4.5 billion years by emitting an alpha particle:

92-uranium-238
↓
90-thorium-234 + alpha particle (nucleus of 2-helium-4)

The mass number of uranium-238 declines by four and its atomic number by two when it emits an alpha particle. The number before the element name is the atomic number and that after the element name is the mass number. The totals of the atomic numbers and the mass numbers, respectively, on both sides of the nuclear reaction must be the same. (This is like balancing a chemical equation, in which the number of atoms of each element on both sides of the reaction must be equal.)

In beta decay, the atomic number increases by one if an electron is emitted or decreases by one if a positron is emitted. For instance, thorium-234, which is the decay product of uranium-238, in turn beta-decays into protactinium-234 by emitting an electron:

90-thorium-234
↓
91-protactinium-234 + beta particle (electron)

The nuclei that result from radioactive decay may themselves be radioactive. Therefore, some radioactive elements have decay chains

---

[257] The terms alpha, beta, and gamma radiation, and X-rays were coined because scientists did not know the nature of these kinds of radiation when they were first detected.

that may contain many radioactive elements, one derived from the other. (See Appendix B for a diagram of the decay chain of uranium-238.)

The radioactive decay of nuclei is described probabilistically. Within any given time period, a particular unstable nucleus has a fixed probability of decay. As a result, each radioactive element is characterized by a "half-life," which is the time it takes for half the initial atoms to decay (or transmute into another element or nuclear state). At the end of one half-life, half the original element is left, while the other half is transformed into another element. After two half-lives, one-fourth of the original element is left; after three half-lives, one-eighth is left, and so on. This results in the build-up of decay products. If the decay products themselves decay into other elements, a number of radioactive materials come into being. The decay products of radioactive elements are also called *daughter products* or *progeny*.

Binding Energy

Nuclei are tightly bound together by the strong nuclear force and each nucleus has a characteristic *binding energy*. This is the amount of energy it would take to completely break up a nucleus and separate all the neutrons and protons in it. Typically, binding energy increases by several megaelectron-volts (MeV) for every proton or neutron added to a nucleus. (Since protons and neutrons are constituent particles of nuclei, they are known collectively as *nucleons*.) The release of nuclear energy derives from the differences in binding energy between the initial nucleus (or nuclei) and relative to the end products of the nuclear reaction, such as fission or fusion.

The electrons that whirl around the nucleus are held together in their orbits by electrical forces. It takes on the order of a few electron-volts to dislodge an electron from the outer shell of an atom. The "binding energy" of a nucleon is on the order of a million times greater. Electrons are the particles the enable chemical reactions; nucleons take part in nuclear reactions. The huge differences in binding energy are one measure of the differences in the quantities of energy derived from nuclear compared to chemical reactions.

It must be stressed that the binding energy is the amount of *energy that would have to be added to the nucleus to break it up*. It can be thought of (approximately) as the amount of energy liberated when a nucleon is drawn into the nucleus due to the short-range nuclear attractive force. Since energy and mass are equivalent, *nuclei with higher binding energy per nucleon have a lower atomic weight per nucleon.*

The key to release of nuclear energy from fission of heavy elements and fusion of light elements is that elements in the middle of the periodic table of elements, with intermediate mass numbers have a higher binding energy per nucleon (that is a lower atomic weight per nucleon). Therefore, when a heavy nucleus is fissioned, the resultant products of the nuclear reaction have a slightly smaller combined nuclear mass. This mass difference is converted to energy during nuclear fission.

Nuclear Fission

Nuclear energy is produced by the conversion of a small amount of the mass of the nucleus of an atom into energy. In principle, all mass and energy are equivalent in a proportion defined by Albert Einstein's famous equation:

$$E = mc^2$$

where E stands for energy, m for mass, and c for the speed of light.

Since the speed of light is a very large number—300 million meters per second—a small amount of mass is equivalent to a very large amount of energy. For instance, one kilogram (about 2.2 pounds) of matter is equivalent to

$$E = 1 \text{ kg} \times (3 \times 10^8 \text{ meters/sec})^2$$
$$= 1 \times 3 \times 10^8 \times 3 \times 10^8 \text{ joules}$$
$$= 9 \times 10^{16} \text{ joules}$$

This is a huge of amount of energy, equivalent to the energy content of roughly three million metric tons of coal.

Heavy atoms, such as uranium or plutonium, can be split by bombarding them with neutrons.[258] The resultant fragments, called fission products, are of intermediate atomic weight, and have a combined mass

---

[258] Nuclear fission can also be induced by bombardment of the nucleus by electrically charged particles, such as alpha particles. However, the nucleus is positively charged and alpha particles are also positively charged. Since positive charges repel each other, these types of fission reactions are more difficult to accomplish than reactions with neutrons. Fission can also be induced by bombarding the nucleus with energetic gamma rays (photons). This process is called photofission.

that is slightly smaller than the original nucleus. The difference appears as energy. As explained in the previous section, this mass difference arises from the binding energy characteristics of heavy elements compared to elements of intermediate atomic weight. Since the binding energy of the fission products per nucleon is higher, their total nucleonic mass is lower. The net result is that fission converts some of the mass of the heavy nucleus into energy.

The energy and mass aspects of the fission process can be explained mathematically as follows. Let the total binding energy of the heavy nucleus and the two fission products be B_h, B_{f1}, and Bf_2, respectively. Then:

Amount of energy released per fission = E_r
$$= (B_{f1} + B_{f2})\ B_h$$

Amount of mass converted to energy = E_r/c^2
$$= \{(B_{f1} + B_{f2}) - B_h\}/c^2$$

This energy appears in various forms: the kinetic energy of the neutrons, the vibrational energy of the fission fragments, and gamma radiation. All of these forms of energy are converted to heat by absorption in with the surrounding media in the reactor, mainly the coolant and the moderator (for thermal reactors).

The most basic fission reaction in nuclear reactors involves the splitting of the nucleus of uranium-235 when it is struck by a neutron. The uranium-235 first absorbs the neutron to yield uranium-236, and most of these U-236 nuclei split into two fission fragments. Fission reactions typically also release two to four neutrons (depending on the speed on the neutrons inducing the fission and probabilistic factors). One of these neutrons must trigger another fission for a sustained chain reaction. The fission reactions in a nuclear reactor can be written generically as follows:

U-235 + n ———> U-236

U-236 ———> fission fragments + 2 to 4 neutrons + 200 MeV energy (approx.)

The uranium-236 nucleus does not split evenly into equal fission fragments. Rather, the tendency, especially with fission induced by thermal neutrons, is for one fragment to be considerably lighter than the other. Figure A-1 shows the distribution of fission products due to fission with the slow neutrons and fast neutrons. It can be seen that the fission product atomic numbers are concentrated in the ranges from about 80 to 105 and from about 130 to 150 in thermal reactors. An example of a fission reaction is:

92-U-235 + n ———> 92-U-236

92-U-236 ———> 38-strontium-90 + 54-xenon-144 + 2 neutrons + energy

While many heavy nuclei can be fissioned with fast neutrons, only a few can be fissioned with "slow" (thermal) neutrons. It turns out that, with some exceptions—e.g., plutonium-240—only nuclei that can be fissioned with slow neutrons can be used for sustaining chain reactions.

Figure A-1: Distribution of Atomic Numbers of Fission Products

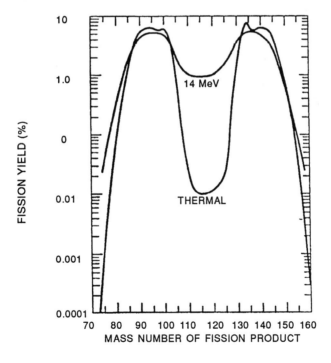

Source: Till and Meyers, eds. 1983, p. 1-5.

Isotopes with nuclei that can be fissioned with zero energy neutrons (in practice neutrons with low energy, or "slow neutrons") are called *fissile* materials. Generally, these are the odd-numbered isotopes, such as uranium-233, uranium-235, plutonium-239, and plutonium-241. Other heavy nuclei, e.g., uranium-238, can be fissioned with fast neutrons and so are *fissionable* but not fissile.

There are only three fissile isotopes of practical importance: uranium-233, uranium-235, and plutonium-239. Of these, only uranium-235 occurs naturally in significant quantities. The other two occur in trace quantities only.

Fertile Materials

To obtain plutonium-239 and uranium-233 in amounts useful for nuclear energy production, they must be manufactured from materials that occur in relative abundance. Plutonium-239 is produced from reactions following the absorption of a neutron by uranium-238; uranium-233 is produced by neutron absorption in thorium-232. Uranium-238 and thorium-232 are called *fertile materials*, and the production of fissile materials from them is called *breeding*.

The reactions for plutonium-239 are:

92-U-238 + n ———> 92-U-239

92-U-239 ———> 93-Np-239 + beta particle (electron)

93-Np-239 ———> 94-Pu-239 + beta particle (electron)

For uranium-233 the reactions are:

90-Th-232 + n ———> 90-Th-233

90-Th-233 ———> 91-Pa-233 + beta particle (electron)

91-Pa-233 ———> 92-U-233 + beta particle.

(The symbol Pa stands for the element protactinium.)

APPENDIX B: URANIUM

First discovered in the nineteeth century, uranium is an element found everywhere on Earth, but mainly in trace quantities. In 1938, German physicists Otto Hahn and Fritz Strassmann showed that uranium could be split into parts to yield energy. Uranium is the principal fuel for nuclear reactors and the main raw material for nuclear weapons.

Natural uranium consists of three isotopes: uranium-238, uranium-235, and uranium-234, as shown in Table B-1. All uranium isotopes are radioactive. Uranium-238, the most prevalent isotope in uranium ore, has a half-life of almost 4.5 billion years; that is, half the atoms in any sample will decay in that amount of time. Uranium-238 decays by alpha emission into thorium-234, which itself decays by beta emission to protactinium-234, which decays by beta emission to uranium-234, and so on. The various decay products, (sometimes referred to as "progeny" or "daughters") form a series starting at uranium-238. After several more alpha and beta decays, the series ends with the stable isotope lead-206. See Figure B-1.

Table B-1: Uranium Isotopes and Natural Uranium Composition

Isotope	Percent in Natural Uranium	Number of Protons	Number of Neutrons	Half-life (Years)
Uranium-238	99.284	92	146	4.46 billion
Uranium-235	0.711	92	143	704 million
Uranium-234	0.005	92	142	256,000

Uranium-238 emits alpha radiation. As noted in Appendix A, this is less penetrating than other forms of radiation. Uranium-238 also emits weak gamma rays. So long as uranium-238 remains outside the body, it poses relatively low health hazard (mainly from the gamma-rays) provided exposure is not prolonged and contact is not close. If inhaled or ingested, however, its radioactivity poses increased risks of lung cancer and bone cancer. Uranium is also chemically toxic at high concentrations and can cause damage to internal organs, notably the kidneys. Animal studies suggest that uranium may affect reproduction, the developing fetus,[259] and increase the risk of leukemia and soft tissue cancers.[260]

Figure B-1: Uranium Decay Chain Main Branch

Uranium-238
(half-life: 4.46 billion years)

⇓ *alpha decay*

Thorium-234
(half-life: 24.1 days)

⇓ *beta decay*

Protactinium-234m
(half-life: 1.17 minutes)

⇓ *beta decay*

Uranium-234
(half-life: 245,000 years)

⇓ *alpha decay*

Thorium-230
(half-life: 75,400 years)

⇓ *alpha decay*

Radium-226
(half-life: 1,600 years)

⇓ *alpha decay*

---

[259] ATSDR 1990.
[260] Filippova et al. 1978

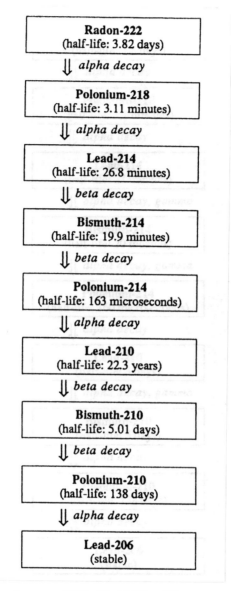

Source: IEER World Wide Web page.

The Mining and Milling Process

Traditionally, uranium has been extracted from open-pits and underground mines. In the past decade, alternative techniques, such as *in-situ* leach mining—in which solutions are injected into underground deposits to dissolve uranium—have become more widely used. Most mines in the U.S. have shut down. Production in 1993 was just under 3,000 metric tons. Annual consumption, averaged over the 1988-1992 period, was about 15,000 metric tons of refined uranium (uranium content).[261]

The milling (refining) process extracts uranium oxide (U_3O_8) from ore to form *yellowcake*, a yellow or brown powder that contains about 90 percent uranium oxide.[262] Conventional mining techniques generate a substantial quantity of mill tailings waste during the milling phase, because the uranium content of the ore is generally less than 1 percent. (*In-situ* leach mining leaves the unusable portion in the ground, and hence does not generate this form of waste). The total volume of mill tailings generated in the U.S. is over 95 percent of the volume of all radioactive waste from all stages of the nuclear weapons and power production.[263] While the specific activity of mill tailings is low relative to many other radioactive wastes, the large volume and lack of an environmental remediation law until 1978 combined to result in widespread contamination. Moreover, the half-lives of the principal radioactive components of mill tailings—thorium-230 and radium-226—are long, being about 75,000 years and 1,600 years, respectively. Contamination is also widespread in other uranium mining countries.[264]

The most serious health hazard associated with uranium mining is lung cancer due to inhaling uranium decay products. Uranium mill tailings contain radioactive materials, notably radium-226, and heavy metals (e.g., manganese and molybdenum) which can leach into groundwater. Near tailings piles, water samples have shown levels of some contaminants at hundreds of times the government's acceptable

---

[261] *Statistical Abstract of the United States 1995*, Table 1201. Note that changes in supplier and utility inventories must be taken into account in calculating annual consumption.

[262] Benedict et al. 1981, p. 265. Note that pure U_3O_8 is black. Yellowcake gets its color from the presence of ammonium diuranate.

[263] Based on the total volume of all radioactive waste (including spent fuel, high-level waste, transuranic waste, low-level waste, and uranium mill tailings) from all sources (both commercial and military) produced in the U.S. since the 1940s, as compiled in Saleska et al. 1989, Appendix C.

[264] See Makhijani et al., eds. 1995, Chapter 5.

level for drinking water.[265]

Mining and milling operations have disproportionately affected indigenous populations around the globe. For example, in the U.S. nearly one-third of all mill tailings from abandoned mill operations are on lands of the Navajo nation alone.[266] Many Native Americans (as well as other workers) have died of lung cancers linked to their work in uranium mines. Others continue to suffer the effects of land and water contamination due to seepage and spills from tailings piles.[267]

Conversion and Enrichment

Uranium is generally used in reactors in the form of uranium dioxide (UO_2) or uranium metal; nuclear weapons use the metallic form. Production of uranium dioxide or metal requires chemical processing of yellowcake. Further, most civilian and many military reactors require uranium that has a higher proportion of uranium-235 than present in natural uranium. The process used to increase the amount of uranium-235 relative to uranium-238 is known as *uranium enrichment.*

U.S. civilian power plants typically use 3 to 5 percent uranium-235. Weapons use "highly enriched uranium" (HEU) with over 90 percent uranium-235. Some research reactors and all U.S. naval reactors also use HEU.

Most uranium enrichment techniques require that uranium first be put in the chemical form uranium hexafluoride (UF_6). After enrichment, UF_6 is chemically converted to uranium dioxide or metal. A major hazard in both the uranium conversion and uranium enrichment processes comes from the handling of uranium hexafluoride, which is chemically toxic as well as radioactive. Moreover, it reacts readily with moisture, releasing highly toxic hydrofluoric acid. Conversion and enrichment facilities have had a number of accidents involving uranium

---

[265] EPA 1983, Vol. 1, pp. D-12-13.

[266] Gilles et al. 1990.

[267] In 1979, a dam holding water in a mill tailings settling pond at the United Nuclear Fuels Corporation mill near Church Rock, New Mexico, gave way and released 94 million gallons of contaminated water into the Puerco River which cuts through Navajo grazing lands. (Makhijani et al., eds. 1995, p. 121.)

hexafluoride.[268]

The bulk of residue from the enrichment process is *depleted* uranium—so-called because most of the uranium-235 has been extracted from it. Depleted uranium has been used by the U.S. military to fabricate armor-piercing conventional weapons and tank armor plating. It was incorporated into these conventional weapons without informing armed forces personnel that depleted uranium is a radioactive material and without procedures for measuring doses to operating personnel.

The enrichment process can also be reversed. Highly enriched uranium can be diluted, or "blended down" with depleted, natural, or very low-enriched uranium to produce 3 to 5 percent low-enriched reactor fuel. Uranium metal at various enrichments must be chemically processed so that it can be blended into a homogeneous material at one enrichment level. As a result, the health and environmental risks of blending are similar to those for uranium conversion and enrichment.[269]

Regulations in the U.S.

In 1983, the federal government set standards for controlling pollution from active and abandoned mill tailings piles resulting from yellowcake production. The principal goals of federal regulations are to limit the seepage of radionuclides and heavy metals into groundwater and reduce emissions of radon-222 to the air.

Until 1997 the U.S. Nuclear Regulatory Commission (NCR) used ad hoc guidelines (developed by its staff in 1981) for decommissioning. In 1997, the NRC published its final decommissioning regulations. The U.S. Environmental Protection Agency (EPA) did not accept these rules for facilities not regulated by the NRC. The EPA's decision was based on the failure of the NRC to explicitly incorporate EPA standards for protecting drinking water into the NRC decommissioning rules.[270]

---

[268] One such accident at the Sequoyah Fuels conversion plant in Gore, Oklahoma, killed one worker and hospitalized 42 others, as well as approximately 100 residents. (Makhijani et al., eds. 1995, pp. 204-205.)

[269] See Makhijani and Makhijani 1995, Chapter 7, for a description of various processes for blending down HEU.

[270] NRC 1997a.

The Future

Uranium and associated decay products thorium-230 and radium-226 will remain hazardous for thousands of years. Current U.S. regulations, however, cover a period of 1,000 years for mill tailings and at most 500 years for "low-level" radioactive waste. This means that future generations—far beyond those promised protection by these regulations—will likely face significant risks from uranium mining, milling, and processing activities.

APPENDIX C: PLUTONIUM

Plutonium-239 is one of the two fissile materials used for the production of nuclear weapons. The other fissile material is uranium-235. Plutonium-239 is virtually non-existent in nature. It is made by bombarding uranium-238 with neutrons in a nuclear reactor. Uranium-238 is present in quantity in most reactor fuel; hence plutonium-239 is continuously made in these reactors. Since plutonium-239 can itself be split by neutrons, it provides a portion of the energy released in a nuclear reactor. The physical properties of plutonium metal are summarized in Table C-1

Table C-1: Physical Characteristics of Plutonium Metal

Color	silver
Melting Point	641 C
Boiling Point	3,232 C
Density	16 to 20 grams per cc

Nuclear Properties of Plutonium

Plutonium belongs to the class of elements called *transuranic elements,* wherein atomic numbers are higher than 92—the atomic number of uranium. Essentially all transuranic materials in existence are man-made. The atomic number of plutonium is 94.

Plutonium has 15 isotopes with mass numbers ranging from 232 to 246. Isotopes of the same element have the same number of protons in their nuclei but differ by the number of neutrons. Since the chemical characteristics of an element are governed by the number of protons in the nucleus, which equals the number of electrons when the atom is electrically neutral (the usual elemental form at room temperature), all isotopes have nearly the same chemical characteristics. This means that in most cases it is very difficult to separate isotopes from each other by chemical techniques.

From the point of view of applications, the two most important plutonium isotopes are plutonium-238 and plutonium-239. Plutonium-238, which is made in nuclear reactors from neptunium-237, is used to produce thermoelectric generators. Plutonium-239 is used for nuclear weapons and for energy. Plutonium-241, although fissile, is impractical both as a nuclear fuel and a material for nuclear warheads. Some of the reasons are far higher cost, shorter half-life, and higher specific activity than plutonium-239. Isotopes of plutonium with mass numbers 238, 240, 241, and 242 are made along with plutonium-239 in nuclear reactors. In this appendix, we focus on civilian and military plutonium, consisting mainly of plutonium-239 mixed with varying amounts of these other isotopes.

Plutonium-239 and plutonium-241 are fissile. Each fission of plutonium-239, resulting from a slow neutron absorption, ends in the production of a little more than two neutrons on the average. The even isotopes—plutonium-238, -240, and -242—are not fissile but yet are fissionable. That is, they can only be split by high energy neutrons. Generally, fissionable but non-fissile isotopes cannot sustain chain reactions; plutonium-240 is an exception to that rule.

The amount of material necessary to achieve a critical mass depends on the geometry and the density of the material, among other factors. The critical mass of a bare sphere of plutonium-239 metal is about ten kilograms. It can be considerably lowered in various ways. The amount of plutonium used in fission weapons is in the three-to-five kilogram range. According to a recent Natural Resources Defense Council report, nuclear weapons with a destructive power of one kiloton can be built with as little as one kilogram of weapon-grade plutonium.[271] The smallest theoretical critical mass of plutonium-239 is only a few hundred grams.

All isotopes of plutonium are radioactive, but they have widely varying half-lives. The half-life is the time it takes for half the atoms of an element to decay. For instance, plutonium-239 has a half-life of about 24,110 years while plutonium-241 has a half-life of 14.4 years. The various isotopes also have different principal decay modes. Table

---

[271] Cochran and Paine 1994. For comparison, the bomb dropped on Nagasaki on August 9, 1945, had about 6.1 kg of plutonium and a destructive power of about 20 kilotons.

C-2 shows a summary of the radiological properties of five plutonium isotopes.

Table C-2: Radiological Properties of Important Plutonium Isotopes

Isotope	Pu-238	Pu-239	Pu-240	Pu-241	Pu-242
Half-life, years	87.74	24,110	6,537	14.4	376,000
Specific activity, curies per gram	17.3	.063	.23	104	.004
Principal decay mode	alpha	alpha	alpha, some spontaneous fission (Note 1)	beta	alpha, some spontaneous fission (Note 1)
Decay energy, MeV	5.593	5.244	5.255	0.021	4.983
Radiological hazards	alpha, weak gamma	alpha, weak gamma	alpha, weak gamma	beta, weak gamma (Note 2)	alpha, weak gamma

Source: CRC 1988. Various sources give slightly different figures for half-lives and energies.

Notes: 1. Source of neutrons causing added radiation to workers in nuclear facilities. A little spontaneous fission occurs in most plutonium isotopes.
2. Plutonium-241 decays into americium-241, which is an intense gamma emitter.

Chemical Properties and Hazards of Plutonium.

Table C-3 describes the chemical properties of plutonium in air. These properties are important because they affect the safety of storage and of operation during processing of plutonium. The oxidation of plutonium represents a health hazard, since the resulting insoluble stable compound, plutonium dioxide is in particulate form that can be easily inhaled. It tends to stay in the lungs for long periods, and is also transported to other parts of the body. Ingestion of plutonium is considerably less dangerous, since very little is absorbed while the rest passes through the digestive system.

Table C-3: How Plutonium Reacts in Air

Forms and Ambient Conditions	*Reaction in Air*
Non-divided metal at room temperature	relatively inert; slowly oxidizes
Divided metal at room temperature	readily reacts to form PuO_2
Fine particles under about one millimeter diameter	spontaneously ignites at about 150 C
Humid, elevated temperatures	readily reacts to form PuO_2

Important Plutonium Compounds and Their Uses

Plutonium combines with oxygen, carbon, and fluorine to form compounds which are used in the nuclear industry, either directly or as intermediates. Some important plutonium compounds are:

- plutonium dioxide (PuO_2), the form used in mixed oxide fuel;

- various carbides (PuC, PuC_2, and Pu_2C_3) which are potential chemical forms for use as fuel;

- nitrates and fluorides, which are intermediates in reprocessing or in the conversion of plutonium into metal or plutonium dioxide.

Plutonium metal is insoluble in nitric acid and plutonium dioxide is slightly soluble in hot, concentrated nitric acid. However, when plutonium dioxide and uranium dioxide form a solid mixture, as in spent fuel from nuclear reactors, then the solubility of plutonium dioxide in nitric acid is enhanced due to the fact that uranium dioxide is soluble in nitric acid. This property is used when reprocessing irradiated nuclear fuels, as for example in the PUREX process.

Formation and Grades of Plutonium-239

As noted in the text, plutonium-239 is formed in both civilian and military reactors from uranium-238. The subsequent absorption of a neutron by plutonium-239 results in the formation of plutonium-240.

Absorption of another neutron by plutonium-240 yields plutonium-241. The higher isotopes are formed in the same way. Since plutonium-239 is the first in a string of plutonium isotopes created from uranium-238 in a reactor, the longer a sample of uranium-238 is irradiated, the greater the percentage of heavier isotopes. Plutonium must be chemically separated from the fission products and remaining uranium in the irradiated reactor fuel. This chemical separation is called *reprocessing*.

Fuel in power reactors is irradiated for longer periods at higher power levels, called high "burn-up," because it is fuel irradiation that generates the heat required for power production. Burn-up is usually measured in "megawatt-days thermal." One megawatt-day thermal is equal to the amount of thermal energy released by fission at a power level of one megawatt thermal operating for one day. If the goal is production of plutonium for military purposes, then the "burn-up" is kept low so that the plutonium-239 produced is as pure as possible—that is, the formation of the higher isotopes, particularly plutonium-240, is kept to a minimum. The burn-up of fuel used to produce weapon-grade plutonium is generally under 1,000 megawatt days per metric ton. Plutonium has been classified into grades by the U.S. DOE (Department of Energy) as shown in Table C-4.

Table C-4: Grades of Plutonium

Plutonium Grade	*Plutonium-240 Content*
Supergrade (see note)	2 to 3 percent
Weapon grade	less than 7 percent
Fuel grade	7 to 19 percent
Reactor grade	more than 19 percent

Note: "Supergrade" is generally blended with lower grades to yield "weapon-grade" plutonium.

It is important to remember that this classification of plutonium according to grades is somewhat arbitrary. For example, although "fuel grade" and "reactor grade" are less suitable as weapons material than "weapon-grade" plutonium, they can also be made into a nuclear weapon, although the yields are less predictable because of unwanted neutrons from spontaneous fission. The ability of countries to build nuclear arsenals from reactor-grade plutonium is not just a theoretical construct. It is a proven fact. During a June 27, 1994, press conference, Secretary of Energy Hazel O' Leary revealed that in 1962 the United States conducted a successful test with "reactor-grade" plutonium. All grades of plutonium can be used as weapons of radiological warfare that involve weapons that disperse radioactivity, but without a nuclear explosion.

GLOSSARY

ABWR	Advanced Boiling Water Reactor: G.E.'s evolutionary light water reactor
ACRS	Advisory Committee on Reactor Safeguards
AEC	Atomic Energy Commission, 1947-1974. Broken up in 1974 into the Energy Research and Development Administration (ERDA) and the NRC. ERDA later became the DOE.
ALMR	Advanced Liquid Metal Reactor: can be used to breed plutonium.
alpha decay	transformation of the nucleus of an atom by emission of a helium-4 nucleus (which consists of two protons and two neutrons)
alpha radiation	helium-4 nuclei emitted by radioactive decay of some isotopes of heavy elements
anti-neutrino	*see* neutrino
atomic number	the number of protons in the nucleus of an element
beta decay	transformation of the nucleus of an atom by emission of an electron or positron
beta radiation	electrons or positrons emitted during the decay of some radioactive materials
binding energy	the energy that is required to separate the nucleons in a nucleus into separate free particles
breeder reactor	a reactor that is designed to produce more fissile material than it consumes; also sometimes called "fast reactor" since most breeder reactors use fast neutrons for sustaining the nuclear chain reaction.

229

breeding	production of fissile materials plutonium-239 and uranium-233 from the fertile materials uranium-238 and thorium-232
Btu	British thermal unit: the amount of energy gained by a pound of water when its temperature is increased by one degree Fahrenheit
burn-up	the amount of energy that has been generated from a unit of nuclear fuel; usually measured in megawatt days thermal per metric ton of initial heavy metal.
BWR	Boiling Water Reactor: a light water reactor that boils the reactor coolant. The resultant steam is used to drive a steam turbine.
CANDU reactor	CANada Deuterium Uranium reactor: a heavy water-moderated power reactor of Canadian design
CFCs	chlorofluorocarbons: compounds that deplete the stratospheric ozone layer
chemical shim	control of the chain reaction in a nuclear reactor by controlling the chemical composition of coolant water; used as a supplement to the employment of control rods.
control rods	rods made out of a neutron-absorbing material that enable control of the chain reaction in a nuclear reactor
curie	a unit of activity of radioactive substances (named after Marie Curie) equivalent to 3.70 x 10^{10} disintegrations per second, approximately the activity produced by one gram of radium-226. Abbreviation is Ci.
decay	*see* radioactive decay
delayed neutrons	neutrons that are not emitted promptly after a fission reaction, but rather after a delay by some of the fission products

deuterium	an isotope of hydrogen with atomic mass of two, having one proton and one neutron in the nucleus; non-radioactive.
DOE	U.S. Department of Energy, created in 1977 by the elevation of ERDA to cabinet status
dual-purpose reactor	a reactor that produces steam for energy use as well as tritium or plutonium for military use
E.C.C.S.	Emergency Core Cooling System
electron	a negatively charged elementary particle
electron capture	capture by the nucleus of an electron from among those surrounding it
electron-volt	a measure of energy used for atomic and nuclear phenomena (abbreviation eV). It is the amount of energy acquired by an electron traveling through one volt of electric potential difference. It is equal to 1.6×10^{-19} joules.
ERDA	Energy Research and Development Administration, created in 1974 from the breakup of the AEC
erg	a unit of energy in the centimeter-gram-second system of units. It is equal to one-ten-millionth part (10^{-7}) of a joule.
eV	electron-volt
exa-	prefix for one million trillion (or 10^{18}). One metric ton of U.S. coal on the average is approximately 25 billion joules. Therefore, one exajoule is equivalent to about 40 million metric tons of U.S. coal.
fast neutron reactor	a reactor that uses fast (energetic) neutrons to sustain the chain reaction
FBR	Fast Breeder Reactor: a fast neutron reactor that generates more fissile material than it uses

fertile material	a material that is not fissile, but which can be converted into a fissile material; uranium-238 and thorium-232 are the principal fertile materials.
fissile material	a material which can fission when it absorbs a low energy (ideally zero energy) neutron. Fissile materials can sustain chain reactions.
fissionable material	a material that can undergo nuclear fission when bombarded by a neutron. Some materials like uranium-238 are fissionable but not fissile because they undergo fission only when bombarded by energetic neutrons.
gamma radiation	high-energy electromagnetic radiation emitted by some nuclei in the process of radioactive decay
giga-	prefix for billion
gigawatt	one billion watts; the approximate electrical capacity of a large nuclear power plant.
graphite-moderated reactor	a reactor in which graphite is used as a moderater
half-life	the amount of time that it takes half of a given quantity of a radioactive element to decay
heavy water	water in which deuterium has replaced ordinary hydrogen; the symbol D is often used for deuterium. The chemical formula for ordinary water is H_2O; that for heavy water is D_2O.
HLW	High-Level Waste: highly radioactive waste left over after the separation of plutonium and uranium from irradiated reactor fuel
HTGR	High-Temperature Gas-Cooled Reactor
HWR	Heavy Water Reactor: a reactor in which heavy water is used as a moderator and coolant

IFR	Integral Fast Reactor: variant of the liquid metal fast breeder reactor design
isotope	a variant of an element that has the same number of protons but a different number of neutrons in the nucleus. Isotopes of elements have the same atomic numbers, but different mass numbers.
joule	a metric unit of energy, equal to one watt of power operating for one second; one kilowatt hour is equivalent to 3.6 million joules.
kilo-	prefix for one thousand
kilowatt	one thousand watts, a common measure for electrical power capacity
kilowatt hour	A unit of energy equal to 3.6 million joules. It is the amount of energy generated by a one-kilowatt source operating for one hour. Abbreviation is kWh when used in reference to electrical energy.
kWhe	kilowatt-hour electrical; kWh is also a common abbreviation.
kWht	kilowatt-hour thermal, equal to 3.6 million joules of thermal (heat) energy. The specification of energy as thermal or electrical is important in electrical generation because only a portion of thermal energy can be converted to electricity.
LMFBR	Liquid Metal Fast Breeder Reactor: a fast breeder reactor cooled by a liquid metal, usually sodium
LWR	Light Water Reactor: a reactor that uses ordinary water, H_2O, as the moderator and coolant; comes in two basic variants. *See* glossary entries for BWR and PWR.
mass number	the number of neutrons plus protons in the nucleus of an element

mega-	prefix for one million
megawatt	one million watts, a common measure of generating capacity for large power plants. When used by itself in the context of electrical generation, it generally refers to electrical generating capacity, and is abbreviated as MW or MWe. The rate of heat generation can also be measured in megawatts, in which case the term megawatts thermal is used, abbreviated as MWt or MWth.
megawatt days	the amount of energy generated by one megawatt of power output over one day. This is used to measure the degree of burn-up of nuclear fuel, and generally refers to thermal energy output extracted from the fuel.
meltdown	the accidental melting of nuclear reactor fuel rods and fuel
MHTGR	Modular High-Temperature Gas-Cooled Reactor: helium-cooled, carbon-moderated reactor
micron	one-millionth of a meter
mill	one-tenth of one U.S. cent. The cost of electrical power is often expressed in terms of mills per kilowatt hour.
moderator	a material used in a nuclear reactor to slow down the fast neutrons emitted in the process of fission
MOX	mixed uranium dioxide and plutonium dioxide fuel
MRS	Monitored Retrievable Storage: a centralized storage facility for spent fuel from nuclear reactors
MTHM	Metric Tons of Heavy Metal: weight of the heavy metal content in reactor fuel

MTIHM	Metric Tons of Initial Heavy Metal: weight of the heavy metal content in reactor fuel when it is first loaded into a reactor
multiplication factor	the number of fission reactions on average caused by a single fission. A multiplication factor greater than one means a reactor is supercritical, equal to one means exactly critical, and less than one means subcritical.
MWe, megawatt electrical	a measure of electrical generating capacity; also written as MW.
MWt, megawatt thermal	a measure of the heat energy generated in a boiler or reactor; also written as MWth.
MWdth	megawatt days thermal
neutrino	a massless, or nearly massless, neutral particle that carries off surplus energy in some nuclear reactions. An anti-neutrino is identical to a neutrino except for the direction of its spin.
neutron	a neutral elementary particle that occurs in the nuclei of elements (except ordinary hydrogen); free neutrons decay into a proton and an electron.
NRC	Nuclear Regulatory Commission, formed in 1974 from the breakup of the AEC
NRC-NAS	National Research Council of the National Academy of Sciences
nuclear fission	the splitting of the nucleus of a heavy element into two lighter nuclei
nuclear fusion	the fusion of two light nuclei, accompanied by the creation of new light nuclei (or one light nucleus and a neutron) and the release of energy
nucleon	proton or neutron occurring in the nucleus of an element
periodic table	a specific arrangement of all elements in a table in ascending order of atomic number

peta-

prefix for one thousand trillion; energy use on a large scale is often measured in petajoules. One metric ton of U.S. coal on the average is approximately 25 billion joules. Therefore, one petajoule is equivalent to about 40,000 metric tons of U.S. coal.

photofission

process by which fission can be induced by bombarding nucleus of an atom with energetic gamma rays

photon

quantum or unit of a gamma ray or other electromagnetic energy

PIUS

Power Inherent Ultimate Safety: advanced light water reactor designed by Asea Brown Boveri

positron

particle identical to an electron, except that it has a positive electrical charge

PRISM

Power Reactor Inherently Safe Module: liquid metal-cooled breeder reactor design by G.E.

prompt critical

the condition of becoming critical with prompt neutrons only

prompt neutrons

neutrons emitted concomitantly with a fission reaction

proton

an elementary particle with a positive charge equal to that of an electron, but which is about 1,836 times heavier than an electron

PUREX process

a chemical process used to separate uranium and plutonium from the fission products in irradiated fuel and from each other. PUREX stands for Plutonium Uranium Extraction.

PWR

Pressurized Water Reactor: a light water reactor that has water under high pressure (primary water) in the reactor which serves as a moderator and coolant. This primary water heats up water in a secondary circuit. Only

	the water in the secondary circuit is converted to steam, while the primary coolant remains in liquid form.
rad	a unit of absorbed radiation dose defined as deposition of 100 ergs of energy per gram of tissue. *See also* erg.
radioactive decay	the spontaneous process by which the unstable nucleus of an element is transferred (transmuted) into another
radionuclide	a radioactive isotope of an element
reactivity	a number that measures whether and by how much a reactor is subcritical or supercritical. A reactivity of zero corresponds to a reactor being exactly critical. Reactivity greater than zero means the reactor is supercritical, while a reactivity less than zero indicates it is subcritical.
RBMK	reactor of the Chernobyl design; the Russian acronym for "reactor high-power boiling channel type."
reactivity insertion	increase of reactivity
Reactor Cavity Cooling System	system that uses natural circulation of air to passively carry away heat in case of malfunction
reactor core	the core of a reactor consisting of the fuel, moderator (in the case of thermal reactors), and coolant
reactor period	time in which reactor power level increases by a factor of approximately 2.7
rem	one rad of radiation dose multiplied by an empirical factor which measures relative biological damage per unit of energy deposition

reprocessing	generic term for chemical separation of plutonium from fission products and remaining uranium in the irradiated reactor fuel
roentgen	a unit of gamma radiation measured by the number of ionizations it causes in air. One roentgen is, for most practical purposes, approximately equal to one rad.
SAFR	Sodium Advanced Fast Reactor: liquid metal-cooled breeder design by Rockwell International
SBWR	Simplified Boiling Water Reactor: advanced light water design by G.E.
SIR	Safe Integral Reactor: advanced pressurized light water design by Combustion Engineering
spent nuclear fuel	nuclear fuel that has been irradiated in a nuclear reactor and has been withdrawn from it; also called irradiated reactor fuel and spent fuel.
spontaneous fission	spontaneous fission of the nucleus of a heavy element without input of any external particle or energy
thermal reactor	a reactor that uses thermal (or slow) neutrons to sustain the chain reaction
transuranic element	an element with atomic number greater than 92, which is the atomic number of uranium
tritium	a radioactive isotope of hydrogen with a half-life of 12.3 years having one proton and two neutrons in its nucleus. Its principal use is in nuclear weapons. It also has research and commercial uses.
uranium enrichment	process used to increase the amount of uranium-235 relative to uranium-238

watt	a metric unit used to measure power, which is the rate of energy generation or consumption. One watt is equal to one joule per second. One horsepower is equal to 746 watts.
watt hour	one watt of power operating for one hour; equivalent to 3,600 joules of energy.
WPPSS	Washington Public Power Supply System: nuclear power project in western U.S., popularly known as "Whoops," which led to the largest utility bond default in history in 1983
yellowcake	yellow and brown powder containing about 90 percent uranium oxide (U_3O_8)
zircaloy	an alloy of zirconium with 1.2 to 1.7 percent tin and smaller quantities of iron, chromium, and nickel used for making the tubes into which the nuclear fuel for light water reactors is inserted (Benedict et al. 1981, pp. 323-324)

REFERENCES

AEC 1948

Atomic Energy Commission, *Report to the U.S. Congress, No. 4*, Washington, D.C.: AEC, 1948.

AEC 1953

Atomic Energy Commission, *Fourteenth Semiannual Report of the Atomic Energy Commission*, Washington, D.C.: AEC, July 1953.

AEC 1957

Atomic Energy Commission, *Theoretical Possibilities and Consequences of Major Accidents in Large Nuclear Power Plants: A Study of Possible Consequences if Certain Assumed Accidents, Theoretically Possible but Highly Improbable, Were to Occur in Large Nuclear Power Plants*, WASH-740, Washington, D.C.: AEC, March 1957.

Albright et al. 1993

David Albright, Frans Berkhout, and William Walker, *World Inventory of Plutonium and Highly Enriched Uranium 1992*, Oxford: Oxford University Press, 1993.

Alexanderson, ed. 1979

E. Pauline Alexanderson, ed., *Fermi-I: New Age for Nuclear Power: A History of the Enrico Fermi Atomic Power Project, the First Large Fast Breeder Reactor Electric Power Plant, and Its Contributions to the Development of a Long-Range Source of Energy*, LaGrange Park, IL: American Nuclear Society, 1979.

Alfven 1972

Hannes Alfven, "Energy and Environment," *Bulletin of the Atomic Scientists*, May 1972, pp. 5-7.

241

Asselstine 1986

James K. Asselstine, Commissioner, U.S. Nuclear Regulatory Commission, Testimony in *Nuclear Reactor Safety*: Hearings before the Subcommittee on Energy Conservation and Power of the Committee on Energy and Commerce, House of Representatives, May 22 and July 16, 1986, Serial No. 99-177, Washington, D.C.: Government Printing Office, 1987.

ATSDR 1990

Agency for Toxic Substances and Disease Registry, *ATSDR Public Health Statement: Uranium*, Atlanta: ATSDR, December 1990.

Baverstock et al. 1992

Keith Baverstock, Bruno Egloff, Aldo Pinchera, Charles Ruchti, Dillwyn Williams, "Thyroid Cancer after Chernobyl," *Nature*, Vol. 359, No. 6390, 3 September 1992, pp. 21-22.

Bechtel 1986

Bechtel National, Inc., *Concept Description Report, Reference Modular High Temperature Gas-Cooled Reactor Plant*, DOE/HTGR-86-118, under subcontract to Gas-Cooled Reactor Associates, for the Department of Energy, San Francisco: Bechtel National, October 1986.

Benedict et al. 1981

Manson Benedict, Thomas H. Pigford, and Hans Wolfgang Levi, *Nuclear Chemical Engineering*, 2nd ed., New York: McGraw-Hill, 1981.

Berkovitz 1989

Dan M. Berkovitz, "Price-Anderson Act: Model Compensation Legislation?—the Sixty-Three Million Dollar Question," *The Harvard Environmental Law Review*, Vol. 13, No. 1, 1989, pp. 1-68.

Bupp and Derian 1978

Irvin C. Bupp and Jean-Claude Derian, *Light Water: How the Nuclear Dream Dissolved*, New York: Basic Books, 1978.

Byrnes et al. 1961   John A. Byrnes, Richard L. McKinley, and Richard C. Legg, "The SL-1 Accident," *Nuclear Engineering*, Vol. 6, No. 58, March 1961, p. 94. (The article in the journal is unsigned; authors' names are from the manuscript.)

Carter 1987   Luther J. Carter, *Nuclear Imperatives and the Public Trust: Dealing with Radioactive Waste*, Washington, D.C.: Resources for the Future, 1987.

Chernousenko 1991   V.M. Chernousenko, *Chernobyl: Insight from the Inside*, Berlin: Springer-Verlag, 1991.

Cochran and Norris 1993   Thomas B. Cochran and Robert S. Norris, *Russian/Soviet Nuclear Warhead Production. Nuclear Weapons Databook Working Papers*, NWD 93-1, Washington, D.C.: Natural Resources Defense Council, 1993.

Cochran and Paine 1994   Thomas B. Cochran and Christopher E. Paine, *The Amount of Plutonium and Highly Enriched Uranium Needed for Pure Fission Nuclear Weapons*, Washington, D.C.: Natural Resources Defense Council, 22 August 1994.

Cochran et al. 1987   Thomas B. Cochran, William M. Arkin, Robert S. Norris, and Milton M. Hoenig, *Nuclear Weapons Databook, Vol. II, U.S. Nuclear Warhead Production*, Cambridge, MA: Ballinger, 1987.

Cochran et al. 1987a   Thomas B. Cochran, William M. Arkin, Robert S. Norris, and Milton M. Hoenig, *Nuclear Weapons Databook, Vol. III, U.S. Nuclear Warhead Facility Profiles*, Cambridge, MA: Ballinger, 1987.

Cohn 1990   Steven Cohn, "The Political Economy of Nuclear Power (1945-1990): The Rise and Fall of an Official Technology," *Journal of Economic Issues*, Vol. XXIV, No. 3, September 1990.

Cohn 1997   Steven M. Cohn, *Too Cheap to Meter: An Economic and Philosophical Analysis of the Nuclear Dream*, Albany: State University of New York Press, 1997.

Cole 1953   Sterling Cole, Letter to Congressman John Phillips, May 20, 1953, with cover note from AEC secretary Roy Snapp, July 9, 1953, DOE Archives, Box 1290, Folder 2.

CRC 1988   *CRC Handbook of Chemistry and Physics, 1988-1989*, Boca Raton, FL: CRC Press, 1988.

Davidson 1950   Ward F. Davidson, "Nuclear Energy for Power Production," *Atomics*, November 1950, pp. 320-327.

Demaree 1970   Allan T. Demaree, "G.E.'s Costly Venture into the Future," *Fortune*, October 1970, pp. 88-93, 156, 158.

DOE 1986   U.S. Department of Energy, *The United States Civilian Nuclear Power Policy, 1954-1984: A Summary History*, DOE/MA-0152, Washington, D.C.: History Division, U.S. Department of Energy, February 1986.

DOE 1991   U.S. Department of Energy, *Draft Environmental Impact Statement for the Siting, Construction, and Operation of New Production Reactor Capacity*, DOE/EIS-0144D, Washington, D.C.: DOE Office of New Production Reactors, April 1991.

DOE 1995   U.S. Department of Energy, *Integrated Data Base Report—1994: U.S. Spent Nuclear Fuel and Radioactive Waste Inventories, Projections, and Characteristics*, DOE/RW-006, Rev. 11, Washington, D.C.: Office of Civilian Radioactive Waste Management, DOE, 1995.

Dubash 1994   Navroz K. Dubash, "Commoditizing Carbon: Social and Environmental Implications of Joint Implementation," in Prodipto Ghosh and Jyotsna Puri, eds., *Joint Implementation of Climate Change Commitments: Opportunities and Apprehensions*. New Delhi: Tata Energy Research Institute, 1994, pp. 51-85.

Duncan 1990   Francis Duncan, *Rickover and the Nuclear Navy, The Discipline of Technology*, Annapolis, MD: Naval Institute Press, 1990.

EIA 1993   Energy Information Administration, *Uranium Purchases Report 1992*, DOE/EIA-0570(92), Washington, D.C.: EIA, August 1993.

EPA 1983   U.S. Environmental Protection Agency, *Final Environmental Impact Statement for Standards for the Control of Byproduct Materials from Uranium Ore Processing*, Washington, D.C.: EPA, 1983.

EPA 1993   U.S. Environmental Protection Agency Science Advisory Board, *Review of the Release of Carbon-14 in Gaseous from High-Level Waste Disposal*, EPA-SAB-RAC-93-010, Washington, D.C.: EPA, 1993.

Falk and Brownlow 1989   Jim Falk and Andrew Brownlow, *The Greenhouse Challenge: What's To Be Done?*, Ringwood, Victoria, Australia: Pengiun, 1989.

Faltermayer 1988   Edmund Faltermayer, "Taking Fear Out of Nuclear Power," *Fortune*, 1 August 1988.

Fermi et al. 1944   Enrico Fermi, James Franck, T.R. Hogness, Zay Jeffries (Chairman), R.S. Mulliken, R.S. Stone, and C.A. Thomas, *Prospectus on Nucleonics* (on microfilm), Washington, D.C.: National Archives Record Group 77, Harrison Bundy Files, 1944. Also known as the "Compton Report."

Filippova et al. 1978   L. G. Filippova, A. P. Nifatov, and E. R. Lyubchanskii, "Some of the long-term sequelae of giving rats enriched uranium," *Radiobiologiya*, Vol. 18, No. 3, 1978, pp. 400-405. (Originally in Russian; translated in NTIS UB/D/120-03 [DOE-TR-4/9], Springfield, VA: National Technical Information Service.)

Ford 1982   Daniel Ford, *The Cult of the Atom: The Secret Papers of the Atomic Energy Commission*, New York: Simon & Schuster, 1982.

Ford Foundation 1974   Ford Foundation Energy Policy Project, *A Time to Choose: America's Energy Future*, Cambridge, MA: Ballinger, 1974.

Fuller 1975   John G. Fuller, *We Almost Lost Detroit*, New York: Reader's Digest Press, 1975.

GA (undated)   General Atomics brochure, "The Modular High Temperature Gas-Cooled Reactor: Inherently Safe Nuclear Power," undated.

GA 1991   F.A. Silady and H. Gotschall, "The 450 MW(t) MHTGR: More Power with the Same High Level of Safety," GA-A20532, (s.l.): General Atomics, August 1991.

Gilles et al. 1990   Cate Gilles, Marti Reed, and Jacques Seronde, *Our Uranium Legacy*, 1990. (Available from Southwest Research and Information Center, Albuquerque, NM).

Golay 1993   Michael W. Golay, "What Role Should Nuclear Power Play, and What Would Life Be Without It?" in *Proceedings of the 2nd MIT International Conference on the Next Generation of Nuclear Power Technology*, October 25-26, 1993, pp. 1-2 to 1-13, MIT-ANP-CP-002, Cambridge, MA: MIT, 1993.

Goldemberg et al. 1988 Jose Goldemberg, Thomas B. Johansson, Amulya K.N. Reddy, and Robert H. Williams, *Energy for a Sustainable World*, New York: Wiley, 1988.

Goodman 1949 Clark Goodman, "Future Developments in Nuclear Power," *Nucleonics*, February 1949.

Goran and Gammill 1963 John R. Goran and William P. Gammill, "The Health Physics Aspects of the SL-1 Accident," *Health Physics*, Vol. 9, 1963, pp. 177-186.

Greenwald 1991 John Greenwald, "Time to Choose," *Time*, April 29, 1991.

Hewlett and Duncan 1990 Richard G. Hewlett and Francis Duncan, *Atomic Shield: A History of the United States Atomic Energy Commission, Volume II, 1947-1952*, Berkeley: University of California Press, 1990.

Hewlett and Holl 1989 Richard G. Hewlett and Jack M. Holl, *Atoms for Peace and War 1953-1961: Eisenhower and the Atomic Energy Commission*, Berkeley: University of California Press, 1989.

IPCC 1996 Intergovernmental Panel on Climate Change, Contribution of Working Group II to Second Assessment Report, "Energy Supply Mitigation Options" (H. Ishitani and T.B. Johansson, eds.), Chapter 19 in R.T. Watson, M.C. Zinyowera, and R. H. Moss, eds., *Climate Change 1995: Impacts, Adaptations, and Mitigation of Climate Change*, Cambridge: Cambridge University Press, 1996.

IPPNW and IEER 1992 International Physicians for the Prevention of Nuclear War and the Institute for Energy and Environmental Research, *Plutonium: Deadly Gold of the Nuclear Age*, Cambridge, MA: International Physicians Press, 1992.

| Jaffe 1981 | Leonard Jaffe, "Technical Aspects and Chronology of the Three Mile Island Accident," in *The Three Miles Island Nuclear Accident: Lesson and Implications, Annals of the New York Academy of Sciences*, Vol. 365, April 4, 1981, pp. 37-47. |

Johansson et al., eds. 1993 — Thomas B. Johansson, Henry Kelly, Amulya K.N. Reddy, and Robert H. Williams, eds., *Renewable Energy: Sources for Fuels and Electricity*, Washington, D.C.: Island Press, 1993.

Kinzig and Kammen 1998 — Ann P. Kinzig and Daniel M. Kammen, "National Trajectories of Carbon Emissions: Analysis of Proposals to Foster the Transition to Low-carbon Economies," *Global Environmental Change*, Vol. 8, No. 3, Fall 1998, pp. 183-208.

Komanoff 1981 — Charles Komanoff, *Power Plant Cost Escalation: Nuclear and Coal Capital Costs, Regulation, and Economics*, New York: Komanoff Energy Associates, 1981.

Komanoff 1985 — Charles Komanoff, "Dismal Science Meets Dismal Subject: The (Mal)practice of Nuclear Power Economics," in *New England Journal of Public Policy*, Fall 1985, pp. 47-59.

Komanoff, Brailove, and Wallach 1997 — C. Komanoff, R. Brailove, and J. Wallach, *Good Money After Bad: An Economic Analysis of the Early Retirement of the Salem Nuclear Generating Station*, White Plains, NY: Pace University School of Law Center for Environmental Legal Studies, September 1997.

Komanoff and
Roelofs 1992

Charles Komanoff and Cora Roelofs, *Fiscal Fission: The Economic Failure of Nuclear Power*, A Greenpeace Report on the Historical Costs of Nuclear Power in the United States, Washington, D.C.: Komanoff Energy Associates, December 1992.

Lamarsh 1983

John R. Lamarsh, *Introduction to Nuclear Engineering*, 2nd ed., Reading, MA: Addison-Wesley Publishing Company, 1983.

Leverett 1947

M.C. Leverett, "Some Engineering Aspects of Nuclear Energy," Clinton Laboratories, Oak Ridge, MDDC-1304, Manhattan Engineer District Declassified papers, 14 August 1947.

Lidsky 1986

Lawrence M. Lidsky, "Modular Gas-Cooled Reactors for Electric Power Generation," presented at International Nuclear Engineering Symposium on the Development and Use of Small- and Medium-Size Power Reactors in the Next Generation, Tokai University, November 19-21, 1986.

Lidsky 1987

Lawrence M. Lidsky, "Safe Nuclear Power and the Coalition Against It," *The New Republic*, December 28, 1987.

Lidsky 1988

Lawrence M. Lidsky, "Nuclear Power: Levels of Safety," *Radiation Research*, Vol. 113, 1988, pp. 217-226.

Lidsky and Cohn
1993

Lawrence M. Lidsky and S.M. Cohn, "What Now? An Examination of the Impact of the Issues Raised," in *Proceedings of the 2nd MIT International Conference on the Next Generation of Nuclear Power Technology*, pp. 4-33 to 14-38, MIT-ANP-CP-002, Cambridge, MA: MIT, 1993.

Likhtarev et al.
1995

I.A. Likhtarev, B.G. Sobolev, I.A. Kairo, N.D. Tronko, T.I. Bogdanova, V.A. Oleinic, E.V. Epshtein, and V. Beral, "Thyroid Cancer in the Ukraine," *Nature*, Vol. 375, No. 6530, June 1995, p. 365.

Lilienthal 1947

David E. Lilienthal, "Atomic Energy and American Industry," *Combustion*, November 1947, pp. 34-37.

Makhijani 1990

Arjun Makhijani, *Draft Power in South Asian Foodgrain Production: Analysis of the Problem and Suggestions for Policy*, Takoma Park, MD: Institute for Energy and Environmental Research, September 1990.

Makhijani 1995

Arjun Makhijani, "Calculating Doses from Disposal of High-Level Radioactive Waste: Review of a National Academy of Sciences Report," *Science for Democratic Action*, Vol. 4. No. 4, Fall 1995.

Makhijani 1995a

Arjun Makhijani, "Comments of Dr. Arjun Makhijani to the Massachusetts Department of Public Utilities on the Notice of Inquiry and Order Seeking Comments on Electric Industry Restructuring," Institute for Energy and Environmental Research, Takoma Park, MD, April 12, 1995.

Makhijani and
Makhijani 1995

Arjun Makhijani and Annie Makhijani, *Fissile Materials in a Glass, Darkly: Technical and Policy Aspects of the Disposition of Plutonium and Highly Enriched Uranium*, Takoma Park, MD: IEER Press, 1995.

Makhijani and
Poole 1975

Arjun Makhijani, in collaboration with Alan Poole, *Energy and Agriculture in the Third World*, Cambridge, MA: Ballinger, 1975.

Makhijani and
Saleska 1992

Arjun Makhijani and Scott Saleska, *High-Level Dollars, Low-Level Sense: A Critique of Present Policy for the Management of Long-Lived Radioactive Wastes and Discussion of an Alternative Approach*, New York: The Apex Press, 1992.

Makhijani and
Zerriffi 1996

Arjun Makhijani and Hisham Zerriffi, "The U.S. Can't Have It Both Ways," *Bulletin of the Atomic Scientists*, March-April 1996, pp. 36-39.

Makhijani et al.,
eds. 1995

Arjun Makhijani, Howard Hu, and Katherine Yih, eds., *Nuclear Wastelands: A Global Guide to Nuclear Weapons Production and Its Health and Environmental Effects*, Cambridge, MA: MIT Press, 1995.

Mann 1955

Martin Mann, "Atomic Power Plant to Serve New York," *Popular Science,* August 1955, pp. 82-85.

May 1989

John May, *The Greenpeace Book of the Nuclear Age*, New York: Pantheon Books, 1989.

Mazuzan and
Walker 1984

George T. Mazuzan and J. Samuel Walker, *Controlling the Atom: The Beginnings of Nuclear Regulation 1946-1962*, Berkeley: University of California Press, 1984.

McWhorter
et al.

see Westinghouse 1995

Medvedev 1990

Zhores A. Medvedev, *The Legacy of Chernobyl*, New York: W. W. Norton, 1990.

MRS Review
Commission 1989

Monitored Retrievable Storage Review Commission, *Nuclear Waste: Is There a Need for Federal Interim Storage?*, Washington, D.C.: Government Printing Office, 1989.

Mullenbach 1963

Philip Mullenbach, *Civilian Nuclear Power: Economic Issues and Policy Formation*, New York: The Twentieth Century Fund, 1963.

Murray 1953a   Thomas E. Murray, Memorandum to Lewis L.
                      Strauss, August 25, 1953, DOE Archives,
                      Record Group 326, Box 1290, Folder 2.

Murray 1953b   Thomas E. Murray, Memorandum to Lewis L.
                      Strauss, September 16, 1953, DOE Archives,
                      Record Group 326, Box 1290, Folder 2.

Murray 1953c   Thomas E. Murray, Memorandum to Lewis L.
                      Strauss, September 18, 1953. DOE Archives,
                      Record Group 326, Box 1290, Folder 2.

NAS 1994   Committee on International Security and
                      Arms Control, *Management and Disposition
                      of Excess Weapons Plutonium*, Washington,
                      D.C.: National Academy Press, 1994.

NAS 1995   Panel on Reactor-Related Options for the
                      Disposition of Excess Weapons Plutonium,
                      Committee on International Security and
                      Arms Control, *Management and Disposition
                      of Excess Weapons Plutonium—Reactor-
                      Related Options*, Washington, D.C.: National
                      Academy Press, 1995.

NEI 1995   Nuclear Energy Institute, *The U.S. Nuclear
                      Industry's Strategic Plan for Building New
                      Nuclear Power Plants: 5th Annual Update*,
                      Washington, D.C.: NEI, November 1995.

NEI 1996   Nuclear Energy Institute, *Nuclear Energy
                      Insight96*, Washington, D.C.: NEI, January
                      1996.

NEI 1998   Nuclear Energy Institute (website),
                      "Advanced-Design Nuclear Power Plants,"
                      http://www.nei.org/public/library/infob6.htm,
                      September 1998.

NPOC 1990   Nuclear Power Oversight Committee,
                      *Strategic Plan for Building New Nuclear
                      Power Plants*, (s.l.): NPOC, November 1990.

NRC 1987   Nuclear Regulatory Commission, *Report on the Accident at the Chernobyl Nuclear Power Station*, NUREG-1250, Washington, D.C.: NRC, January 1987.

NRC 1995   Nuclear Regulatory Commission, *Information Digest: 1995*, NUREG-1350, Vol. 7, Washington, D.C.: NRC, 1995.

NRC 1997   Nuclear Regulatory Commission, *Information Digest: 1997*, NUREG-1350, Vol. 9, Washington, D.C.: NRC, 1977.

NRC 1997a   Nuclear Regulatory Commission, "Radiological Criteria for License Termination," *Federal Register*, Vol. 62, No. 139, July 21, 1997, pp. 39058-39092.

NRC-NAS 1983   Waste Isolation Systems Panel, *A Study of the Isolation System for Geologic Disposal of Radioactive Wastes*, Washington, D.C.: National Academy Press, 1983.

NRC-NAS 1995   Committee on the Technical Bases for Yucca Mountain Standards, *Technical Bases for Yucca Mountain Standards*, Washington, D.C.: National Academy Press, 1995.

NRC-NAS 1996   Committee on Separations Technology and Transmutation Systems, *Nuclear Wastes: Technologies for Separations and Transmutation*, Washington, D.C.: National Academy Press, 1996.

Nucleonics 1953   "Nuclear Power Feasibility Studies," *Nucleonics*, Vol. 11, No. 6, June 1953, pp. 49-64.

Nucleonics 1953a   Editorial, "It's in the Lap of Congress," *Nucleonics*, A special report, Vol. 11, No. 9, September 1953.

Nucleonics 1953b   Editorial, "Pros and Cons of the PWR," *Nucleonics*, Vol. 11, No. 12, December 1953.

Nucleonics 1958	"PWR: The Significance of Shippingport, A Nucleonics Special Report," *Nucleonics*, Vol. 16, No. 4, April 1958, pp. 53-72.
Nucleonics Week 1989	"Outlook on Advanced Reactors," *Nucleonics Week*, March 30, 1989, pp. 1-20.
Okrent 1981	David Okrent, *Nuclear Reactor Safety: On the History of the Regulatory Process*, Madison: University of Wisconsin Press, 1981.
Ott 1953	H.C. Ott, Memorandum to J.A. Lane, USAEC, May 25, 1953, with cover note from AEC general manager M.W. Boyer, June 5, 1953, DOE Archives, Record Group 326, Box 1290, Folder 2.
Paley Commission 1952	The President's Materials Policy Commission (Paley Commission), *Resources for Freedom*, 5 vols., Washington, D.C.: Government Printing Office, 1952.
Perry et al. 1977	Robert Perry, A.J. Alexander, W. Allen, P. deLeon, A. Gandara, W.E. Mooz, E. Rolph, S. Siegel, and K.A. Solomon, *Development and Commercialization of the Light Water Reactor, 1946-1976*, R-2180-NSF, Santa Monica, CA: Rand, June 1977.
Pigford 1995	Thomas H. Pigford, *The Yucca Mountain Standard: Proposals for Leniency*, UCB-NE-9525, Berkeley: University of California Department of Nuclear Engineering, November 1995. (Also in *Proceedings of the Materials Research Society, Scientific Basis of Nuclear Waste Management*, November 1995.)
Pigford 1996	Thomas H. Pigford, "Historical Aspects of Nuclear Energy Utilization in the Half-Century and Its Prospect Toward the 21st Century," *Journal of Nuclear Science and Technology*, Vol. 33, No. 3, Japan, 1996.

Price-Anderson Act 1957	*Public Law 85-256*, September 2, 1957, commonly called the Price-Anderson Act of 1957, an amendment to the *Atomic Energy Act of 1954, Public Law 83-703*.
Pringle and Spigelman 1981	Peter Pringle and James Spigelman, *The Nuclear Barons*, New York: Holt, Rinehart & Winston, 1981.
Sachs 1995	Noah Sachs, *Risky Relapse into Reprocessing: Environmental and Non-Proliferation Consequences of the Department of Energy's Spent Fuel Management Program*, Takoma Park, MD: Institute for Energy and Environmental Research, 1995.
Saleska et al. 1989	Scott Saleska, et al. *Nuclear Legacy: An Overview of the Places, Politics, and Problems of Radioactive Waste in the United States*, Washington, D.C.: Public Citizen, 1989.
Schurr and Marschak 1950	Sam H. Schurr and Jacob Marschak, *Economic Aspects of Atomic Power: An Exploratory Study,*, Princeton: Princeton University Press, 1950.
Shearson Lehman 1993	Shearson Lehman Brothers, *Electric Utilities Commentary*, Vol. 3, No. 1, January 6, 1993.
Silady and Gotschall	*see* GA 1991
Sporn 1950	Philip Sporn, "Prospects in Industrial Application of Atomic Energy," *Bulletin of the Atomic Scientists*, October 1950, pp. 303-306, 320.
Sporn 1951	Philip Sporn, "Development of Atomic Energy—How Can Private Industry Best Participate?" *Nucleonics*, Vol. 8, No. 2, February 1951.

Starr 1953   Chauncey Starr, "The Role of Multipurpose Reactors," *Nucleonics*, January 1953, pp. 62-64.

Statistical 1994/1995   *Statistical Abstract of the United States 1994* and *1995*, Washington, D.C.: Bureau of the Census, U.S. Department of Commerce, 1994 and 1995.

Strauss 1954   Lewis L. Strauss, Chairman, U.S. Atomic Energy Commission, "Remarks Prepared for Delivery at the Founders' Day Dinner, National Association of Science Writers," September 16, 1954.

Strauss 1955   Lewis Strauss, Memorandum for General Fields, U.S. AEC, May 25, 1955, Enclosure "a" to cover note from AEC Secretary W.B. McCool, June 10, 1955, DOE Archives, Record Group 326, Box 1290, Folder 2.

Suits 1951   C.G. Suits, "Power from the Atom—An Appraisal," *Nucleonics*, Vol. 8. No. 2, February 1951.

Till and Meyer, eds. 1983   J.E. Till and H.R. Meyer, eds., *Radiological Assessment*, NUREG/CR-3332, Washington, D.C.: U.S. Nuclear Regulatory Commission, 1983.

TMI Commission 1979   President's Commission on the Accident at Three Mile Island, *Report of the President's Commission on the Accident at Three Mile Island: The Accident at Three Mile Island*, Washington, D.C.: The Commission. 1979.

Todd and Stall 1997   D. M. Todd and H. Stall, "Integrated Gasification Combined-Cycle: The Preferred Power Technology for a Variety of Applications," GE Power Systems, Schenectady, Paper presented at the Power-Gen Europe 97 Conference, Madrid, June 1997.

UCS 1990

Union of Concerned Scientists, *Advanced Reactor Study*, prepared by MHB Technical Associates, Cambridge, MA: UCS, July 1990.

von Hippel et al. 1986

Frank von Hippel, David H. Albright, and Barbara G. Levi, *Quantities of Fissile Materials in US and Soviet Nuclear Weapons*, PU/CEES Report No. 168, Princeton: Center for Energy and Environmental Studies, Princeton University, 1986.

WASH-740

see AEC 1957

Weaver 1959

Charles H. Weaver, Letter to John A. McCone, Chairman of the AEC, November 7, 1959. (Copy from the National Archives in Washington, D.C.)

Weinberg 1972

Alvin Weinberg, "The Safety of Nuclear Power," Presentation to the Council for the Advancement of Science Writing, Briefing on New Horizons in Science, Boulder, Colorado, November 14, 1972.

Weinberg 1994

Alvin M. Weinberg, *The First Nuclear Era: The Life and Times of a Technological Fixer*, New York: American Institute of Physics Press, 1994.

Westinghouse 1995

D.L. McWhorter, R.L. Geddes, W.N. Jackson, and W.C. Bugher, *Chemical Stabilization of Defense Related and Commercial Spent Fuel at the Savannah River Site*, NMP-PLS-950239, Aiken, SC: Westinghouse Savannah River Company, 16 August 1995.

Williams 1993

Robert H. Williams, "The Outlook for Renewable Energy," *Proceedings of the Second MIT International Conference on the Next Generation of Nuclear Power Technology*, pp. 4-11 to 4-32, and discussion on pp. 4-39 to 4-46, MIT-ANP-CP-002, Cambridge, MA: MIT, 1993.

Williams and
Cantelon, eds.
1984

Robert C. Williams and Philip L. Cantelon, eds., *The American Atom: A Documentary History of Nuclear Policies from the Discovery of Fission to the Present, 1939-1984*, Philadelphia: University of Pennsylvania Press, 1984.

Yergin 1991

Daniel Yergin, *The Prize: The Epic Quest for Oil, Money & Power*, New York: Simon & Schuster, 1991.

Zerriffi 1996

Hisham Zerriffi, *Tritium: The Environmental, Health, Budgetary, and Strategic Effects of the Department of Energy's Decision to Produce Tritium*, Takoma Park, MD: Institute for Energy and Environmental Research, 1996.

INDEX

Advisory Committee on Reactor Safeguards (ACRS), 4, 89, 96-97, 100-01, 145
Alpha decay, 208, 215
Alpha emission, 215
Alpha particle, 209
Alpha radiation, 34, 215
American Association for the Advancement of Science, 62
Anti-neutrino. *See* Neutrino.
Argonne National Laboratory, 49, 69
Asselstine, James, 6, 135
Atom, 223
 bomb, 9, 23, 54, 57
 nucleus of, 207-13, 222
 peaceful, 59, 61
 structure of, 207
"Atoms for Peace," 3, 12, 57, 60-62, 79
Atomic age, 2, 54
Atomic Energy Act, 57, 69, 84
Atomic Energy Commission (AEC), 1-4, 7, 53-55, 57, 59, 64-65, 68-69, 71-72, 74-79, 81, 84-87, 89-90, 96-98, 100-101, 106
 response to safety problems, 6
 split-up of, 99
Atomic mass, 207
Atomic number, 207-09, 213, 222
"Atomic pile," 21
Atomic power
 economics of, 63
 peaceful applications of, 3, 59-62
Atomic weight, 211-12

Bechtel, 2, 67
Beta decay, 208-09, 215

Beta particle, 209
Boiling water reactor (BWR), 39, 86, 88, 92, 109
 Advanced (ABWR), 138
 Simplified (SBWR), 139
Brookhaven National Laboratory, 5, 89-90

CANada Deuterium Uranium reactor (CANDU), 40, 44, 91
Cancer risk, 34, 160-62, 216, 218-19
Carbon, 28, 40
 emissions, 127, 206
 pyrolytic, 142
Carbon dioxide, 28, 40, 187
 emissions, 1, 10-11, 133, 178, 180, 185, 190, 195, 197, 199-200, 204-06
Chain reaction, 20, 23, 42, 54, 92
 creating, 24
 sustaining, 21-22, 30, 36-38, 40, 212-13, 223
Chernobyl, 5-6, 10-11, 13, 40, 42, 72, 91-92, 94, 134-35, 137, 143, 146, 151, 153-64, 178
"China Syndrome," 93
Chlorofluorocarbons (CFCs), 178, 193
Clinton Laboratories, 64
Coal, 17, 20, 22, 32, 146, 187, 192, 211
 price of, 28-29, 63, 65, 70
Cohn, Steve, 86
Cold War, 3-4, 12, 53, 57, 61, 70, 75-76, 81, 83, 151, 165-66, 173, 180
Cole, Sterling, 59,

Combined–cycle natural gas plants, 11, 195, 197-200
Combustion Engineering, 138-39
Commercial-military link, 10, 165, 173, 176
Commonwealth Edison, 2, 67, 86
Congress, U.S., 2, 57, 65, 89-90, 99, 101, 126-28, 135, 168
Consolidated Edison Company, 2, 66, 86
Control rods, 42-43, 91-92, 144
Coolant, 36-39, 44, 48, 92-95, 105, 109, 140-46, 212
  breeder reactor, 49
  radioactive, 40
Core meltdown accidents, 8, 89, 97, 100-01, 135, 138-39, 141, 143-44, 146, 148-49, 151
Criticality, 40, 42, 114
  prompt, 43, 91, 114, 144
Critical mass, 21, 34, 223
Cult of the Atom, The, 54, 96

Daughter product. See Progeny.
Davidson, Ward, 2, 66-67
Davis, Kenneth, 84
"Decay heat," 93
Department of Energy (DOE), 9-10, 12, 14, 49, 71, 75, 99, 107, 122, 124, 126-28, 133, 141, 144, 147, 168, 171, 173, 175, 226
Detroit Edison, 2, 67-68, 100
Deuterium. See Hydrogen, heavy.
Dietz, David, 17
"Doubling time," 45
Dow Chemical, 2, 67-68
Duquesne Light Company, 61, 77, 79, 86

Einstein, Albert, 20, 211
Eisenhower, Dwight D., 3, 57-58, 60-61, 79
Electricity, 182, 191-94
  coal-fired, 106, 185
  cost, 104, 189-92, 199
  generation, 26-29, 64, 72, 82, 84, 93, 147, 166, 168, 174, 176-77, 179, 184-85, 187, 191, 197, 206
  nuclear, 3, 50, 63, 68, 77, 99, 107, 121, 177-79, 185, 197
  oil-generated, 105
  production, 141, 178, 180, 200
Electricity capital fund, 12, 193
Electron, 207-08, 210, 222
  capture, 208-09
  emission, 208-09
Elements
  characteristics of, 222
  heavy, 22, 208, 211-12
  light, 20, 211
  non-radioactive, 172
  radioactive, 209-10
  transuranic, 222
Emergency Core Cooling System (E.C.C.S.), 94, 97-98, 139
Energy, 54 , 70, 73, 99, 148, 163, 181, 211-12, 215, 222
  atomic, 57, 59, 68-69, 82
  binding, 210-12
  chemical, 26
  cost, 104, 141, 166, 184, 189, 191
  efficiency, 11-12, 182-85, 189-90, 193-94, 199
  electrical, 53, 60
  fission, 20, 38
  generation, 33
  global, 99, 176, 179-80, 184-85
  kinetic, 22, 38, 212
  mechanical, 26-27
  production, 36, 42, 44-45, 48, 178
  solar, 4, 26, 82-83, 194
  sources, renewable, 11-12, 82-83, 183-84, 187-89, 193-95, 199
  supply, 17, 23, 99, 176, 181-83, 200
  thermal, 36, 38, 148, 185, 194, 197, 226
  vibrational, 36, 212
  wind, 82, 187, 194
"Energy for Peace" program, 12
Energy Research and Development

Agency (ERDA), 99
Environment, 133, 143, 191-92
  concerns about, 99, 108, 118-20,
    124-25, 140, 165, 168, 171,
    184, 193, 199
  despoliation of, 18, 63, 91, 94,
    181-82
  mismanagement of, 8
Environmental Protection Agency
  (EPA), 127, 220
Experimental Breeder Reactor I, 4,
  99-101, 151

Fallout, 154, 158-60
Fermi, Enrico, 2, 65, 100
Fermi I reactor, 4, 100-101, 144
Fissile material, 4, 13, 22-23, 76,
  114, 214, 222-23
Fission, 20, 28, 40, 124, 210-11,
  213, 223
  fragments, 36, 212
  nuclear, 24, 36, 42, 73, 82, 109,
    211
  process, 43, 108, 212
  products, 8, 24, 33-36, 38, 50-51,
    66, 93, 99, 101, 108-09, 115,
    143, 145-46, 169, 171, 211-
    13, 226
  reactions, 63, 66, 143, 212-13
  spontaneous, 208-09, 227
Fissionable materials, 22, 60, 214,
  223
Force
  attractive, 210-11
  electrical, 210
  strong nuclear, 210
Ford, Daniel, 54, 96
Forrestal, James, 55
Fortune magazine, 87, 133
Fossil fuels, 24, 73, 86, 105, 121,
  133, 178, 180, 201, 205
  as an energy source, 23, 26, 104
  large-scale use of, 1, 11-12, 18,
    28, 179
Fuel "kernels," 141-42, 146
Fuel rods, 36, 42, 92-93, 95, 98

spent, 109
Fusion, 210-11

Gamma rays, 36, 45, 160, 208-09,
  212, 215-16
General Atomics, 141, 147, 172
General Electric (G.E.), 2, 7, 39,
  62, 78, 84-88, 99-100, 138-40
Golay, Michael, 6
Government subsidies, 7, 9, 67, 72,
  84, 86, 88, 103, 168, 171
Greenhouse gases
  build-up of, 3, 10-11, 134, 178-79
  reduction of, 10-11, 121, 180,
    187, 192, 194-95, 200-01, 203-
    06

Hafstad, Lawrence, 65
Hahn, Otto, 215
Half-life, 210, 218
Heat engine, 26, 37
Heavy water-moderated reactor
  (HWR), 40, 44-45
Helium, 34, 40, 140-42, 146, 172,
  208-09
*High-Level Dollars, Low-Level
  Sense,* 13
Hiroshima, bombing of, 23, 57, 159
Hutchins, Robert M., 54
Hydrogen, 38, 49, 98, 146, 165,
  171, 187
  bomb, 53, 58-59
  explosion, 95-96
  heavy, 40, 49

Idaho National Engineering Labora-
  tory, 98
"Inherently safe" reactors, 6, 133-49
Institute for Energy and Environ-
  mental Research, 124, 165
Integral Fast Reactor (IFR), 49
*Introduction to Nuclear Engineer-
  ing,* 36
Isotopes, 22
  fissile, 22, 30, 37, 45, 50, 115,
    214

plutonium, 34, 37, 45, 115, 223-
    24, 226
    radioactive, 50, 96, 165, 171,
        208, 215, 223
    separation of, 222-23
    stable, 215

Jackson, C.D., 58
Jersey Central Power & Light, 88
Joint Committee on Atomic Energy,
    congressional, 3, 56, 59, 75, 78-
    79, 85, 97
"Just-in-time" system, 12, 195

Kyoto Protocol, 180, 200-05

Lamarsh, John R., 36
Leverett, M.C., 64
Lidsky, Lawrence, 92, 142, 147
Light water reactor (LWR), 3, 5-7,
    32, 39-40, 43-45, 52, 62, 72, 75,
    84, 88, 91-93, 95, 97, 103, 109,
    122, 135, 137-40, 142-44, 149-
    51, 163, 166, 171
    advanced (ALWR), 136, 138-39
    core meltdowns in, 4, 6, 89, 134
    safety vulnerabilities of, 86, 91,
        96, 136
    spent fuel, 114-15
Lilienthal, David, 64, 69
Liquid metal fast breeder reactor
    (LMFBR), 49, 140, 143-44, 150
"Loose nukes," 173
Loss-of-coolant accidents, 5-6, 38,
    89, 91-99, 138-39, 143, 149, 151

Manhattan Project, the, 23, 34, 58,
    63, 78, 100, 115, 151
Massachusetts Institute of Technol-
    ogy (MIT), 6, 68, 92, 142, 147
Mass number, 207-09, 211, 222-23
Medvedev, Zhores, 155, 157-58,
    160-61
Moderator, 37-39, 44, 48, 92, 142,
    146, 212
    solid, 40

Modular High-Temperature Gas-
    Cooled Reactor (MHTGR), 137,
    140-47, 171-72
Monitored Retrievable Storage
    (MRS), 107, 117-19, 128
Monsanto Chemical Company, 2,
    64, 67, 71-72
MOX (mixed-oxide) fuel, 10, 12,
    50, 122-23, 166, 170-71, 175, 225
Mullenbach, Philip, 68, 72
Multiplication factor, 40, 42, 143
Murray, Thomas, 3, 59, 79

Nagasaki, bombing of, 23, 159
National Research Council of the
    National Academy of Sciences
    (NRC-NAS), 126-27, 165, 169
National Science Foundation, 88
Native American land, 14, 107,
    110, 117-18, 126, 128, 219
Natural gas, 18, 28, 187, 193-94,
    197-200
    availability, 105
    technologies, 11
Natural gas combined–cycle plants.
    See Combined-cycle natural gas
    plants.
Nautilus, 79
Navy, U.S., 44, 69-70, 73-76, 78-
    81, 85, 100
Neutrino, anti-, 207
Neutron, 22, 24, 100, 142-43, 210,
    212, 222, 227
    absorption, 40, 42-43, 66, 214,
        223, 225-26
    bombardment, 2, 33, 37, 66, 172,
        211, 222
    decay, 207
    delayed, 43, 91
    free, 207
    high speed, 36-38, 43, 48, 213-
        14, 223
    prompt, 43
    slow, 37, 48, 212-14, 223
Non-proliferation issues, 8, 10, 37,
    118-20, 122, 124-26, 171, 173

Nuclear accidents, 5-6, 10-11, 13,
24, 50, 89-95, 97-98, 100, 134,
139, 144, 146, 149-51, 155, 164,
168, 179
preventing, 67, 142, 162-63, 178
Nuclear electricity, 3, 50, 63, 68,
77, 99, 107, 121, 177-79, 185
Nuclear energy, 28-29, 44-45, 51,
53-54, 64, 78, 81-84, 151, 176,
179, 185, 210-11
commercial, 20-21, 23, 57, 195
military applications, 20-21, 23,
57, 64-65, 107
peaceful uses of, 57, 59-60
secrecy surrounding, 69
widespread use of, 22, 62, 64,
166
Nuclear Energy Institute, 133
Nuclear fuel, 9-10, 13, 24, 30, 82-
83, 223, 225-26
cost, 63-66, 68
disposal of, 9, 14, 24, 29, 113,
118-19, 121, 123-25, 128, 175
fissile, 33
management of, 8-9, 107, 116,
121, 123, 180
refractory coated, 141
spent, 33, 107-09, 113-25, 128,
140, 168, 170, 175, 180, 225
storage, 117-19, 125
Nuclear Non-Proliferation Treaty
(NPT), 173, 176, 180
Nuclear power, 30, 35, 51, 53, 69
commercial, 12, 63-64, 70-71, 84-
85, 88, 99, 101, 150, 165, 175-
76
cost of, 1, 7-8, 27, 29, 57, 63,
65, 67-68, 80, 85-86, 88, 90,
101-06, 195
economic viability of, 1-2, 7, 61,
63, 67, 70-72, 75, 81, 87,
103, 134
growth, 9, 56, 60-61, 70, 79-80,
87, 100, 118-21, 149-50, 218
large-scale utilization of, 36, 85
problems facing, 66, 105-06,

119, 179-80
proliferation vulnerability of, 8,
120, 179, 199
promotion of, 133, 136, 165
resistance to, 135, 141, 148, 187,
192
"unwarranted optimism" about,
2, 65
world generation of, 178-79
Nuclear Power Oversight Committee, 133, 136
Nuclear power plants, 36, 66, 69,
72, 84, 108, 118-21, 123, 143,
150, 178, 195, 198
commercial, 3, 75, 77, 85, 109,
115, 133-34, 140, 165
cost, 64, 68, 88, 101-06, 134,
138, 172, 191, 199
dangers of, 4, 96, 102, 135, 155,
164, 180
"Great Bandwagon Market" for,
87-88
safety requirements for, 103, 163
vs. coal-fired plants, 27-28, 53,
62-64, 68, 79, 81, 84-86, 104,
106, 177, 180, 187, 197, 199-
200
Nuclear reactors, 22-23, 30, 32-36,
38, 40, 50, 54, 65-66, 71, 106,
137, 148, 173, 175, 212, 215, 222
design of, 37, 66-67, 94, 172
economic viability of, 69, 175
government-supported, 70, 86,
88, 103, 165
safety, 89, 96, 103, 134, 141,
147, 163
Nuclear Regulatory Commission
(NRC), 6, 92, 99, 103, 106, 119,
134-35, 138-39, 144-47, 220
Nuclear waste, 9, 106, 108-09, 113,
115, 119-20, 123, 125, 127-28,
168
volume of, 110, 128
Nuclear waste fund, 9, 107, 119,
123
Nuclear Waste Policy Act, 107,

116, 120, 126
Nuclear weapons, 22-23, 30, 33, 57-
    58, 108, 113-14, 125, 128, 165,
    168-73, 215, 223, 226-27
    establishment, secrecy in, 13, 101
    preventing proliferation of, 24, 60
    production, 70-71, 98-99, 178-
        79, 218-19, 222, 227
Nuclear weapons states, 5, 9, 125,
    176
    arsenals of, 8, 180
Nucleon, 210, 212
Nucleonics, 62, 70

Oak Ridge National Laboratory, 15,
    64, 78, 98, 128, 144, 148
Oil, 18, 28, 182, 192-93
    resources, control of, 19-20
    shortages, 4, 82-83, 104-05, 189
Okrent, David, 97
Oppenheimer, J.R., 2, 58, 65
Orth, Karlheinz, 137
Ott, H.C., 3, 75
Ozone layer, protecting, 178, 193

Pacific Gas and Electric, 2, 67
Paley Commission, 4, 82
Passive safety systems, 136-49
Pearl Harbor, 19
Periodic table, 208, 211
Petroleum. *See* Oil.
Photon, 209
"Planetary engineering," 54
Plutonium, 30, 50, 52, 54, 99, 108,
    114, 120, 180, 211, 227
    "breeding," 44
    commercial, 10, 67, 77, 115, 125,
        223
    compounds, 225
    cost of, 36, 168, 171, 175
    dangers of, 34, 224
    emissions, 1
    extracting, 34
    military disposition of, 141, 165-
        66, 168-69, 171, 174-75
    military production of, 2, 10, 35,

44, 67, 71-72, 77, 80, 115,
    125, 165, 173, 223, 226
    nuclear weapons-usable, 9, 63,
        68, 71-72, 115, 122, 125, 128,
        168, 170, 223, 226
    plutonium-238, 223
    plutonium-239, 23, 30, 33-34,
        37, 43-45, 48, 71, 99, 115,
        214, 222-23, 225-26
    plutonium-240, 34, 45, 213, 223,
        225-26
    plutonium-241, 34, 45, 115, 214,
        223, 226
    plutonium-242, 34, 223
    processing, 34, 166, 169, 175,
        224
    production, 23-24, 69, 140, 168-
        69, 175
    reprocessing, 52, 122, 168-69,
        225
    surplus, 10, 12, 122, 125, 166,
        169, 174
Pollution, 18, 184, 204
    control regulations, 19, 220-21
    respiratory diseases due to, 19
Positron, 208-09
Power Reactor Inherently Safe Mod-
    ule (PRISM), 140
Pressurized water reactor (PWR), 3-
    4, 39-40, 73-76, 79, 86, 92, 100,
    105, 109, 137-38, 140, 144, 175
    Advanced Passive, 139
    Safe Integral Reactor (SIR), 139
Price-Anderson Act, 5, 7, 13, 84,
    88-90
Process Inherent Ultimate Safety
    (PIUS) reactor, 139
Progeny, 210, 215
Proliferation issues, 9-10, 24, 52,
    140, 165, 168, 170-71, 179
Propaganda, 3-4, 15, 25, 59, 61,
    70, 73-76, 115, 140, 151
"Propaganda capital," 4, 59, 83
Proton, 207, 210, 222

Radiation, 34, 67, 126-28, 208

disease due to, 154-62, 168, 216,
218-19
electromagnetic, 36, 209, 212
gamma, 36, 45, 160, 208-09, 212
tan, 161
weapons, 108, 227
Radioactivity, 14, 24, 29, 66, 93,
99, 108-09, 161, 208, 216, 220
off-site releases of, 6, 37, 135,
156, 159
releases of, 8, 90-91, 94, 96-97,
101, 138, 142-44, 154-55, 159
Radionuclide, 220
migration, retardation of, 14
long-lived, 156
short-lived, 111, 113
Radium, 30, 218, 221
RAND corporation, 88
Reactivity, 40, 42-43, 101, 144-48
Reactor, 34
breeder, 37, 44-45, 48-49, 52, 68,
77-78, 91, 99-100, 140, 144,
166, 171
carbon-moderated, 40
commercial, 5, 8, 10, 12, 37, 68,
73-74, 84, 91, 100, 219, 225
control, 40, 42-44, 91, 101
converter, 45
designs, various, 63, 133, 135,
149-50
development, 3-4, 75-77
dual-use, 10, 12, 67, 71-72, 77
fast, 37, 44, 48
fuel, 31, 33, 35, 37, 44, 50, 52
gas-cooled, 27, 135, 140, 150
graphite-moderated, 5-6, 44, 72,
75, 79, 91, 140, 142, 146,
160, 163
multi-purpose, 69
naval, 44, 69-70, 73-76, 78, 81
passive, 136-38
RBMK, 40, 153, 163
safety, 13, 50, 89-90, 93, 97, 145-
50
sodium-cooled, 4-5, 48-50, 78,
99-100, 135, 150, 163

thermal, 37, 44, 50, 213
"triple-play," 13, 166, 168, 173
water-cooled, water-moderated,
5, 48, 73, 150
Reactor Safeguards Committee, 89
Reprocessing, 10, 14, 34, 52, 78,
108, 120, 125, 128, 140, 166, 175
commercial, 9-10, 35, 118, 121-25
cost of, 9, 35, 122-23
definition of, 9, 226
electrolytic, 49
government-subsidized, 9
military, 125
waste, 108, 124
Reprocessing plants, 50, 124
military, 9, 115
Resources
biomass, 185-87
exploitation of, 19, 181
low-grade, 26
natural, 18
wars for control of, 19
Rickover, Admiral Hyman G., 73-
76, 78-81

Sachs, Noah, 124
Safety, 6-8, 10
criteria, 10-11
requirements, retroactive, 6
systems, 7
Seaborg, Glenn, 2, 54, 65
Seawolf, 78, 100
Shippingport reactor, 76-77, 79-80,
84, 86, 149
Sodium, 48
liquid, 49-50, 99, 135, 140, 150
Sodium Advanced Fast Reactor
(SAFR), 140
Sodium-cooled breeder reactor, 4-5,
99-100
Sporn, Philip, 65, 106
Starr, Chauncey, 68
Strassmann, Fritz, 215
Strauss, Lewis, 1-2, 15, 28-29, 53,
55, 59, 61, 76, 85, 106
Subcriticality, 40, 42

Suits, C.G., 2, 62-63
Supercriticality, 40, 42, 143, 153
Szilard, Leo, 20

Tailings ponds, 31, 218-20
Thomas, Charles, 64, 71
Thorium, 30, 44-45, 48, 50, 142,
   209, 214-15, 218, 221
Three Mile Island accident, 8, 38,
   92, 94-96, 98-99, 102-04, 133-
   34, 137, 163
"Too cheap to meter," 28-29, 53,
   61-62, 84
"Triple-play" reactors, 13
Tritium, 141
   commercial, 13, 173
   military, 12-13, 171-72, 175
   production, 165-66, 168, 171-73,
      175
Truman, Harry S., 82
Turbine, steam, 26-28, 39, 92-93,
   105, 140, 198

Union of Concerned Scientists, 5,
   96, 98, 144-45, 148
Uranium, 9, 15, 28, 50, 52, 59, 78,
   108, 211, 215
   atoms, splitting, 20
   cost, 4, 36, 65, 86
   depleted, 32, 110, 128, 166, 220
   dioxide, 92, 219, 225
   dissolving, 218
   enriched, 32, 44, 75-76, 142, 165-
      66, 172, 219-20
   extraction, 32
   fission, 20-21
   irradiated, 35, 226
   low-enriched, 109, 114, 166, 220
   mills, 30-31, 50, 110, 128, 217-
      19, 221
   natural, 22, 24, 30-32, 37, 40,
      44, 75, 114, 166, 215, 219-20
   ore, 32, 215
   processing, 32, 110, 128, 166, 221
   proliferation, 4
   refined, 218

   residual, 8, 23, 33-34, 115
   safety and, 4, 216
   supposed scarcity of, 4, 36, 63
   uranium-232, 44-45
   uranium-233, 30, 43-45, 48, 142,
      214
   uranium-234, 215
   uranium-235, 4, 22-24, 30-34, 36-
      37, 44-45, 48, 50, 75, 114-15,
      212, 214-15, 220, 222
   uranium-236, 212
   uranium-238, 22-23, 30, 33-34,
      36-37, 44-45, 48, 50, 71, 99,
      115, 209-10, 214-15, 222, 225-
      26

Volcker, Paul, 104-05

Washington Public Power Supply
   System (WPPSS), 105
WASH-740, 5, 89-90, 96-97
Waste management, 9, 13-14, 34,
   108, 111, 113, 115, 119-24, 128
Weaver, Charles H., 80
Weinberg, Alvin, 15, 128
Wells, H.G., 20
Westinghouse, 7, 10, 73-74, 76-80,
   84-88, 122, 124, 138-39, 175
Williams, Robert H., 191
World War I, 19
World War II, 19, 23, 50, 53-54,
   63, 73, 78, 81

Yellowcake, 31, 218, 220
Yucca Mountain repository, 9, 14,
   126-28